SpringerBriefs in Intelligent Systems

Artificial Intelligence, Multiagent Systems,
and Cognitive Robotics

Series Editors
Gerhard Weiss, Maastricht University, Maastricht, The Netherlands
Karl Tuyls, University of Liverpool, Liverpool, UK
    Google DeepMind, London, UK

Editorial Board Members
Felix Brandt, Technische Universität München, Munich, Germany
Wolfram Burgard, Albert-Ludwigs-Universität Freiburg, Freiburg, Germany
Marco Dorigo ⓘ, Université Libre de Bruxelles, Brussels, Belgium
Peter Flach, University of Bristol, Bristol, UK
Brian Gerkey, Open Source Robotics Foundation, Mountain View, USA
Nicholas R. Jennings, Imperial College London, London, UK
Michael Luck, King's College London, London, UK
Simon Parsons, City University of New York, New York, USA
Henri Prade, IRIT, Toulouse, France
Jeffrey S. Rosenschein, Hebrew University of Jerusalem, Jerusalem, Israel
Francesca Rossi, University of Padova, Padua, Italy
Carles Sierra, IIIA-CSIC Cerdanyola, Barcelona, Spain
Milind Tambe, University of Southern California, Los Angeles, USA
Makoto Yokoo, Kyushu University, Fukuoka, Japan

This series covers the entire research and application spectrum of intelligent systems, including artificial intelligence, multiagent systems, and cognitive robotics. Typical texts for publication in the series include, but are not limited to, state-of-the-art reviews, tutorials, summaries, introductions, surveys, and in-depth case and application studies of established or emerging fields and topics in the realm of computational intelligent systems. Essays exploring philosophical and societal issues raised by intelligent systems are also very welcome.

Francesco Leofante • Matthew Wicker

Robust Explainable AI

Francesco Leofante
Department of Computing
Imperial College London
London, UK

Matthew Wicker
Department of Computing
Imperial College London
London, UK

ISSN 2196-548X ISSN 2196-5498 (electronic)
SpringerBriefs in Intelligent Systems
ISBN 978-3-031-89021-5 ISBN 978-3-031-89022-2 (eBook)
https://doi.org/10.1007/978-3-031-89022-2

© The Editor(s) (if applicable) and The Author(s), under exclusive license to Springer Nature Switzerland AG 2025

This work is subject to copyright. All rights are solely and exclusively licensed by the Publisher, whether the whole or part of the material is concerned, specifically the rights of translation, reprinting, reuse of illustrations, recitation, broadcasting, reproduction on microfilms or in any other physical way, and transmission or information storage and retrieval, electronic adaptation, computer software, or by similar or dissimilar methodology now known or hereafter developed.
The use of general descriptive names, registered names, trademarks, service marks, etc. in this publication does not imply, even in the absence of a specific statement, that such names are exempt from the relevant protective laws and regulations and therefore free for general use.
The publisher, the authors and the editors are safe to assume that the advice and information in this book are believed to be true and accurate at the date of publication. Neither the publisher nor the authors or the editors give a warranty, expressed or implied, with respect to the material contained herein or for any errors or omissions that may have been made. The publisher remains neutral with regard to jurisdictional claims in published maps and institutional affiliations.

This Springer imprint is published by the registered company Springer Nature Switzerland AG
The registered company address is: Gewerbestrasse 11, 6330 Cham, Switzerland

If disposing of this product, please recycle the paper.

Foreword

In an era where artificial intelligence is everywhere in our lives, the need for explainable artificial intelligence (explainable AI) has never been more critical. One reason why explainable AI is such a hot topic right now is the over emphasis of evaluating accuracy of machine learning models, at the expense of other important aspects such as understandability, robustness, appropriate trust, and usefulness. In a similar way, explainable machine learning research has perhaps over emphasized explanation correctness at the expense of other important measures. This book, "Robust Explainable AI" by Francesco Leofante and Matthew Wicker, two up-and-coming leaders in explainable AI, aims to address this issue by demystifying the key issues of robust explainability, providing a comprehensive overview of the topic for researchers and practitioners in machine learning.

The book introduces a series of nice examples for why robustness is important. To borrow one from the introduction, consider twins, Alice and Bob, who each apply to the same school and have the same grades, same parents, etc. Both are rejected, but give different reasons for the rejection; for example, Alice is told her parents' income is insufficient and needs to be $10,000$ higher, while Bob is told it is his grades, requiring an A in Science instead of B. This is an example of brittleness in explainable AI—surely both should receive the same explanation when all of their details are the same! Following on, consider if their parents' income increases by $15,000$, and they re-apply, only to be told that their income is not sufficient again. This is brittleness at work again.

Chapter 1 successfully argues the importance and fundamental problems of robustness, noting that a key aspect of the usability and potential success of explainable AI is robustness; while Chap. 2 presents an overview of the concepts in machine learning and explainable machine learning that are required to understand the remainder of the book.

Chapter 3 focuses on robustness of counterfactual goals, highlighting the four main challenges for robustness: (1) Changes to inputs, such as when Alice's parents' income increased; (2) Changes to a model, such as when a new model deploying an updated data is deployed; (3) Model multiplicity, such as where multiple machine learning models are used and their outputs aggregated, but counterfactual explanations for each model disagree; and (4) Noisy execution, where decision subjects, such as Bob, Alice, and their parents, are given advice but cannot execute it exactly; for example, they increase their income by $15,000$ when the explanation recommended increasing it by $10,000$. Yes, this really can happen, as the authors show! They highlight methods for both generating and evaluation the robustness of counterfactual explanations. The chapter ends with an excellent overview of the major challenges that remain in this field and identifies links between robust counterfactual explanation research and adjacent fields, such as fairness, privacy, and model robustness.

Chapter 4 discussed the robustness of saliency explanations for images. Saliency explanations are explanations that highlight the regions of images that were important for a decision, using techniques such as heat maps. Saliency explanations present some overlapping problems with counterfactual explanations, but also provide new challenges due to the input being image data. Not only are the modes of input so different—pixels rather than variables with values—but also "attacks" on model inputs, where minor changes to images radically change the model output and the explanations, are easier to "hide" in images than in tabular data. The chapter provides an overview of several techniques for saliency explanations, discusses how to estimate robustness using sampling techniques, and how to certify the robustness, which is much stronger than simply evaluating via sampling or other example-based approaches. Again, the chapter ends with a nice overview of the main challenge in the field.

The book is filled with a nice balance of technical rigor, for those of us inclined to get into such details, as well as intuitive examples and figures to get the main ideas, for those of us wanting to understand the main concepts and challenges, leaving the technical details for another day.

"Robust Explainable AI" is not only a technical manual that introduces the reader to the field, it also provides a forward-looking perspective of the major challenges and future research directions, making this essential reading for practitioners and researchers interested in explainable AI.

Brisbane, Australia
December 2024

Tim Miller

Preface

Machine Learning (ML) has become a valuable tool for supporting automated decision-making in a wide range of applications. Unfortunately, ML models typically operate in a black-box fashion and often fail to expose the reasoning needed to support the decisions they produce. In some cases, ML algorithms operate in a more transparent way, but follow decision-making processes that are not designed to be communicated to humans. Ultimately, this opacity can undermine trust and hinder deployment of ML in safety-critical or consequential decision-making tasks.

These challenges have spurred a significant increase in research on Explainable AI (XAI), aiming to deliver novel explanation methods that can help shed some light on the outputs of ML models. Despite several advancements in the field, recent work has demonstrated that many state-of-the-art XAI algorithms often lack robustness and may thus produce completely different explanations for similar inputs. This lack of robustness may jeopardize the explanatory function of XAI algorithms and undermine trust in the underlying ML system, rather than engendering it.

Introducing Robust Explainable AI, this book presents innovative approaches to mitigate the limitations of existing XAI algorithms and bolster ML trustworthiness. The book focuses on two mainstream classes of explanation methods: counterfactual explanations and saliency-based explanations. For each explanation type, we discuss the fundamental principles and primary methods for computing them. We then formally define different perspectives on robustness, and cover existing methodologies for generating provably explanations.

The book's exposition is designed to give readers a solid technical understanding of the concepts at hand, while also presenting some open challenges and future directions for research in robustness of each explanation method discussed.

London, UK
September 2024

Francesco Leofante
Matthew Wicker

Acknowledgments

Francesco Leofante would like to thank Nico Potyka and Jay Jiang for providing valuable feedback and comments on earlier versions of this book.

Francesco Leofante was supported by Imperial College London under the Imperial College Research Fellowship scheme.

Competing Interests The authors have no competing interests to declare that are relevant to the content of this manuscript.

Contents

1	**Introduction**		1
	References		2
2	**Explainability in Machine Learning: Preliminaries and Overview**		5
	2.1	Preliminaries and Technical Background	5
		2.1.1 Supervised Machine Learning	5
		2.1.2 Deep Learning	7
		2.1.3 Basics of Explainability	13
	2.2	The Wider World of Explainability	14
	References		15
3	**Robustness of Counterfactual Explanations**		17
	3.1	Introduction	17
	3.2	Counterfactual Explanations	18
	3.3	Robustness of Counterfactual Explanations	21
		3.3.1 Robustness Against Input Changes	22
		3.3.2 Robustness Against Model Changes	25
		3.3.3 Robustness Against Model Multiplicity	29
		3.3.4 Robustness Against Noisy Execution	33
	3.4	Discussion and Future Work	35
	References		36
4	**Robustness of Saliency-Based Explanations**		41
	4.1	Introduction	41
	4.2	Saliency-Based Explanations	41
		4.2.1 The Gradient as an Explanation	43
		4.2.2 Perturbation-Based Explanations	46
		4.2.3 Layer-wise Relevance Propagation	47
	4.3	Metrics for Saliency-Based Explanations	48
		4.3.1 Fidelity Metrics	49
		4.3.2 Localisation Metrics	49
	4.4	Robustness of Saliency-Based Explanations	50

		4.4.1 Defining Robustness for Saliency-Based Explanations	51
	4.5	Estimating Robustness with Random Noise	56
		4.5.1 Estimating Input Robustness via Sampling	56
		4.5.2 Estimating Model Robustness via Sampling	57
		4.5.3 Discussion of the Sampling Approach	58
	4.6	Auditing Saliency-Based Explanations with Counter Examples	58
		4.6.1 Input Attacks on Explanation Robustness	59
		4.6.2 Model Attacks on Explanation Robustness	60
	4.7	Proving Robustness of Saliency-Based Explanations	61
		4.7.1 Certification via Valid Gradient Bounds	62
		4.7.2 Computation of Valid Gradient Bounds	64
		4.7.3 Discussion of Certification	65
	4.8	Literature Review	65
	4.9	Discussion and Future Work	67
References			68

Chapter 1
Introduction

Artificial Intelligence (AI) techniques are increasingly being used for high-stakes decision-making, with applications ranging from autonomous driving [1] to finance [2] and healthcare [3]. Although very powerful, many AI models often produce outputs that are not intelligible to humans, hindering their applicability in consequential decision-making tasks [4]. This lack of transparency can have severe consequences. For instance, in the United States cases have been reported of people being incorrectly denied parole by an AI system [5]. Being able to understand the outputs of AI systems has thus become crucial to ensure their safe deployment in any area where explainability and accountability are paramount.

In response to these concerns, the area of Explainable AI (XAI) has emerged over recent years. XAI is focused on providing methods and tools to explain decisions produced by AI-based systems, with particular emphasis on Machine Learning (ML) models that are built from data. Among these, Deep Neural Networks (DNNs) have been the subject of intense research due to their popularity. Yet, explaining DNNs is far from trivial as these models operate as black boxes, providing no insights on their decision-making processes.

A plethora of approaches have been proposed to compute different kinds of explanations for DNNs and other ML models—see, e.g., [6] for a survey. Among these, two complementary classes of explanation methods have gained prominence: *feature attribution explanations* and *counterfactual explanations*. To understand what makes these explanations advantageous, consider a simple scenario where a bank uses a DNN to predict whether a customer is eligible for a loan. Seeing their application rejected, a customer may want to query the system to understand the reasons behind its decision. A feature attribution explanation would state "*Your application was rejected because the loan amount requested was too high and your credit score was too low*", thus informing the customer about which features in their application influenced the final rejection decision. Instead, a counterfactual explanation for the same decision might state "*If a slightly lower amount were asked,*

© The Author(s), under exclusive license to Springer Nature Switzerland AG 2025
F. Leofante, M. Wicker, *Robust Explainable AI*, SpringerBriefs in Intelligent Systems, https://doi.org/10.1007/978-3-031-89022-2_1

the application would have been accepted", thus pointing to what the customer would need to change in their application for the loan to be granted.

Both of these forms of explainability solve the problem of conveying information used during decision-making to affected users with the intended effect of engendering trust in the underlying system. The usefulness of this information for this goal, however, depends on many factors. We argue that chief among the factors impacting the utility of an explanation to users is its *robustness*. As an initial example for why this is the case, consider a parent applying for a college loan for their twin children, Bob and Alice, who would like to attend the same school (thus the principle of the loan is identical). Now, consider that both loan requests were rejected and the explanation we have received for Bob's loan rejection is "*A key negative factor in our decision was your income and a key positive factor was Bob's age*". Yet, for Alice we received the following "*A key negative factor in our decision was Alice's age and a key positive factor was your income*". The effect of assigning these completely different explanations to such similar applicants is that the underlying system may come across as entirely random and useless or potentially malicious and discriminatory: both undermine trust rather than engendering it. While the example we have given here may seem a caricature of a *worst-case* scenario, researchers have found many such scenarios in real-world machine learning models [7, 8]. In fact, despite many recent advances, state-of-the-art algorithmic solutions *often* fail to produce explanations that are robust, which may jeopardise their function when small changes occur in the setting these explanations were initially generated for. This book aims at introducing Robust Explainable AI, a rapidly growing field that offers novel solutions to alleviate this problem and improve the trustworthiness of explanations.

The rest of this book is organised as follows. Chapter 2 introduces background notions on commonly used ML architectures, as well as general concepts and definitions in Explainable AI. In Chap. 3, we focus on counterfactual explanations and analyse what a lack of robustness means in this context. We discuss four settings in which counterfactual explanations may lack robustness, as well as existing solutions proposed to mitigate this problem, and open problems in this area. Then, in Chap. 4, we provide a comprehensive overview of robustness evaluation for saliency-based explanations, covering fundamental principles, methods and metrics for assessing robustness from various perspectives.

References

1. Khan Muhammad, Amin Ullah, Jaime Lloret, Javier Del Ser, and Victor Hugo C de Albuquerque. Deep learning for safe autonomous driving: Current challenges and future directions. *IEEE Transactions on Intelligent Transportation Systems*, 22(7):4316–4336, 2020.
2. Longbing Cao. AI in finance: challenges, techniques, and opportunities. *ACM Computing Surveys (CSUR)*, 55(3):1–38, 2022.
3. Mohammed Yousef Shaheen. Applications of artificial intelligence (ai) in healthcare: A review. *ScienceOpen Preprints*, 2021.

References

4. Tim Miller. Explanation in artificial intelligence: Insights from the social sciences. *Artif. Intell.*, 267:1–38, 2019.
5. Rebecca Wexler. When a computer program keeps you in jail. *Then New York Times*, 2017. Available at: https://www.nytimes.com/2017/06/13/opinion/how-computers-are-harming-criminal-justice.html (Accessed: August 20th, 2024).
6. Riccardo Guidotti, Anna Monreale, Salvatore Ruggieri, Franco Turini, Fosca Giannotti, and Dino Pedreschi. A survey of methods for explaining black box models. *ACM Comput. Surv.*, 51(5):93:1–93:42, 2019.
7. Ann-Kathrin Dombrowski, Maximillian Alber, Christopher Anders, Marcel Ackermann, Klaus-Robert Müller, and Pan Kessel. Explanations can be manipulated and geometry is to blame. *Advances in Neural Information Processing Systems*, 32, 2019.
8. Junqi Jiang, Francesco Leofante, Antonio Rago, and Francesca Toni. Recourse under model multiplicity via argumentative ensembling. In *Proceedings of the 23rd International Conference on Autonomous Agents and Multiagent Systems, AAMAS 2024, Auckland, New Zealand, May 6–10, 2024*, pages 954–963. International Foundation for Autonomous Agents and Multiagent Systems / ACM, 2024.

Chapter 2
Explainability in Machine Learning: Preliminaries and Overview

2.1 Preliminaries and Technical Background

In this chapter, we cover background topics that are foundational to the concepts discussed in this book. While covering these topics, this chapter establishes a consistent notation. Though non-comprehensive, specifically useful technical background not covered in this section will be discussed at the outset of the chapter in which it is most useful.

This book focuses particularly on deep learning models applied to both tabular and image data. The general principles of explainability that we apply to these settings can be adapted to other such models albeit with varying degrees of difficulty.

2.1.1 Supervised Machine Learning

Throughout this book we will focus primarily on *supervised* learning scenarios in which we assume a joint probability distribution between features $\mathbf{x} \in \mathbb{R}^{n_x}$ and outputs $\mathbf{y} \in \mathbb{R}^{n_y}$ for regression or labels $\mathbf{y} \in \mathbb{Z}^{n_y}$ for classification. We denote the density of this joint distribution with $P(X, Y)$. An observation from this joint distribution is a tuple containing a feature vector (which we will also call an input) and its corresponding label i.e., $(\mathbf{x}, \mathbf{y}) \sim P(X, Y)$. A dataset $\mathcal{D}$ is then simply a collection of such samples, which we index using parenthetical superscripts: $\mathcal{D} := \{(\mathbf{x}^{(i)}, \mathbf{y}^{(i)})\}_{i=1}^{n_\mathcal{D}}$. The goal of a discriminative machine learning algorithm in the supervised learning scenario is then to model the function conditional distribution $P(Y = \mathbf{y} | X = \mathbf{x})$. In non-probabilistic terms, this can be equivalently modelled as learning to fit a label generating function $g : \mathbb{R}^{n_x} \to \mathbb{R}^{n_y}$ (or $g : \mathbb{R}^{n_x} \to \mathbb{Z}^{n_y}$ for classification). How the joint distribution or label generating function are approximated is the subject of supervised machine learning and details depend on the precise *setting*. Below we discuss the two most popular settings: classification and regression.

2.1.1.1 Classification

When **y** comes from a finite set of mutually exclusive values (called classes), the supervised machine learning task is termed *classification*. Where there are c-many classes we denote the set of labels with the integers from 1 to c i.e., $[c] := \{1, 2, \ldots, c\}$ and use n_c to denote its cardinality. It will be useful throughout this book to consider the *one-hot-encoded* versions of our labels which is a vector containing a zero in all but the jth position where $j \in [c]$ is the desired class. To denote the one-hot-encoded labels we use the standard basis notation: $e_{\mathbf{y}^{(i)}}$. Examples of classification tasks include medical diagnosis, loan approval, object recognition. Given an input **x**, a probabilistic machine learning model outputs a multinoulli distribution that represents the predicted conditional distribution $P(Y = \mathbf{y}|X = \mathbf{x})$. The deterministic interpretation of the multinoulli is as "class confidence" scores. Given the parameters of the multinoulli are given as $\hat{\mathbf{y}} \in \mathbb{R}^{n_c}$ (such that $1 = \sum_i \hat{\mathbf{y}}_i$), the likelihood of observing our dataset according to the model that assigns the predictions is: $P(\mathcal{D}) = \prod_i \hat{\mathbf{y}}_{e_{\mathbf{y}^{(i)}}}$. Maximising this likelihood is identical to minimising what is known as the categorical cross-entropy loss which given a $\mathbf{y} \in [c]$ and $\hat{\mathbf{y}}$ is defined as: $-\log(\hat{\mathbf{y}}_{\mathbf{y}})$. We touch on how this is minimised in Sect. 2.1.2.2.

For now, we avoid generalisations to multi-label classification, where a given input may be simultaneously labelled as more than one class. Similarly, we do not discuss the particulars of ordinal regression which is a case where though classes are independent they are mutually related e.g., predicting a students letter grade or the disease stage.

2.1.1.2 Regression

When **y** is a continuous variable, the supervised machine learning task is termed *regression*. Examples of regression tasks include predicting house prices, forecasting stock prices, and predicting steering angles in autonomous driving. Given an input **x**, a probabilistic machine learning model outputs a conditional distribution $P(Y = y|X = \mathbf{x})$ over the possible values of y. For practical applications, this conditional distribution is assumed to be a Gaussian (normal) distribution, parameterised by a mean $\hat{y} \in \mathbb{R}$ and a variance σ^2. The mean $\hat{y}$ represents the predicted value, while the variance σ^2 captures the uncertainty associated with the prediction. The likelihood of observing our dataset $\mathcal{D}$ according to the model, assuming Gaussian noise, is given by:

$$P(\mathcal{D}) = \prod_i \frac{1}{\sqrt{2\pi\sigma^2}} \exp\left(-\frac{(y^{(i)} - \hat{y}^{(i)})^2}{2\sigma^2}\right)$$

2.1 Preliminaries and Technical Background

Maximising this likelihood is identical to minimising what is known as the mean squared error (MSE) loss, which, given a target value y and a prediction $\hat{y}$, is defined as:

$$\text{MSE}(\mathbf{y}, \hat{\mathbf{y}}) = \frac{1}{n}\sum_{i=1}^{n}(y^{(i)} - \hat{y}^{(i)})^2$$

For now, we avoid generalizations to more complex regression tasks such as quantile regression, where the goal is to predict a specific quantile of the target variable, or heteroscedastic regression, where the model accounts for varying levels of noise in the data. Similarly, we do not discuss the particulars of non-parametric regression techniques, such as kernel regression or regression trees, which do not assume a specific functional form for the relationship between the input and the output.

2.1.2 Deep Learning

Throughout this book we will focus primarily on deep learning models i.e., neural networks. In general, we denote a neural network as a parameterised function f^θ whose output is taken as our prediction i.e., $\hat{y} = f^\theta(\mathbf{x})$. The specific details of parameters contained θ depend on the architecture which we discuss in Sect. 2.1.2.1.

2.1.2.1 Architectures

In this section we describe some of the most fundamental elements of popular neural network architectures. We start with the fully connected architecture whose layers will also be employed in the subsequent, more complex architectures we discuss, including convolutional and recurrent architectures.

Fully Connected Neural Networks

Also known as multi-layer perceptrons (MLPs), fully connected neural networks are perhaps the simplest neural network architecture. A single neuron, η can be seen as a parameterised function that takes in an input vector $\mathbf{z} \in \mathbb{R}^m$, a parameter vector $\mathbf{v} \in \mathbb{R}^m$, and a bias $b \in \mathbb{R}$ and returns the output of a non-linearity applied to the affine transformation of the input by the parameters:

$$\eta(\mathbf{z}) = \sigma(\mathbf{v}^\top \mathbf{z} + b)$$

where the non-linear function σ is termed the *activation function*. In order to form a network of neurons, fully-connected neural networks first build a *layer* of neurons which can be described as a list or vector of neurons that share an input and an activation function i.e., $\eta_i(\mathbf{z})$ for $i \in [k]$ where k is termed the *width* of the layer. To

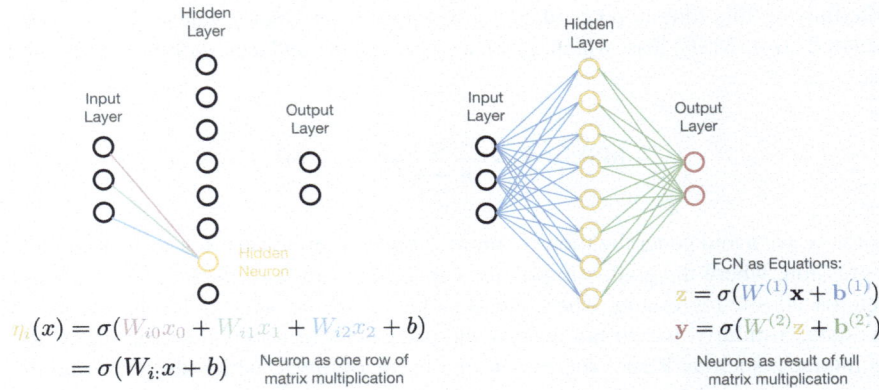

Fig. 2.1 Colour-coded diagram of a fully-connected neural network model. On the left, we depict a colour-coded representation of a single neuron with each input-weight combination having a unique colour that matches its representation in the equation. On the right, we depict a colour-coded representation of a full neural network with each portion of the equation corresponding to the colour of the section in the diagram

ease notation, the parameters of the layer (**v**) and b for each neuron) are collected into a weight matrix $\mathbf{W} \in \mathbb{R}^{m \times k}$ and bias vector $\mathbf{b} \in \mathbb{R}^k$. Finally, the output of this layer can be written as: $\sigma(\mathbf{Wz} + \mathbf{b})$ (Fig. 2.1).

The final step in creating the fully-connected architecture is to chain layers of neurons together. Given that we would like L-many layers we first denote the input as $\mathbf{z}^{(1)} := \mathbf{x}$ and then the *forward pass* through the neural network is given by the following iterative equations:

$$\zeta^{(i)} = \mathbf{W}^{(i)}\mathbf{z} + \mathbf{b}^{(i)} \tag{2.1}$$

$$\mathbf{z}^{(i+1)} = \sigma(\zeta^{(i+1)}), \tag{2.2}$$

where $\mathbf{z}^L$ is defined as the output of the neural network and will often be denoted $\hat{\mathbf{y}}$ following the classical statistical notation for a prediction or estimate. To complete our discussion of the terminology for these models, the output of Eq. (2.1), $\zeta^{(i)}$ is termed the *pre-activation* of the layer with $\mathbf{z}^{(i+1)}$ being the *post-activation* of the layer.

Convolutional Neural Networks

The core idea behind the construction of the convolutional layer is to exploit known spatial structure within the feature dimensions of an input. A prime example of this is the domain of natural images where edges and textures of an image all exist locally around one another. Any network which utilises convolutional layers is to be considered a *convolutional neural network*. Below, we develop the convolutional layer and then briefly discuss the wider literature on convolutional architectures.

2.1 Preliminaries and Technical Background

Fig. 2.2 A diagram of a single convolutions filter acting on a single channel input with color-coded equation for demonstration purposes. The filter we display is two by two (general: $F \times F$) and the weight contributions (in blue) are multiplied by the input values (in black) and are summed over (red and green)

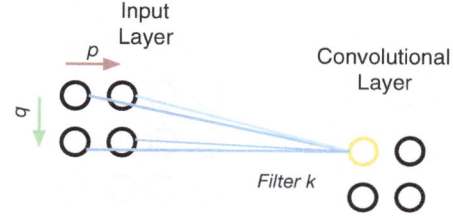

$$z_{ijk} = \sigma \left(\sum_{p=1}^{F} \sum_{q=1}^{F} x_{i+p-1, j+q-1} \cdot W_{pqk} + b_k \right)$$

Formally, given an input $\mathbf{x} \in \mathbb{R}^{H \times W \times C}$ (height H, width W, and C channels) and a set of K filters $\mathbf{W}_k \in \mathbb{R}^{F \times F \times C}$, the output of a convolutional layer $\mathbf{z} \in \mathbb{R}^{H' \times W' \times K}$ is computed as:

$$z_{ijk} = \sigma \left(\sum_{c=1}^{C} \sum_{p=1}^{F} \sum_{q=1}^{F} x_{i+p-1, j+q-1, c} \cdot W_{pqck} + b_k \right) \quad (2.3)$$

where σ is a non-linear activation function, and b_k is a bias term. We visualise a version of the convolution operation above in Fig. 2.2 where we also colour-code portions of the equation. We highlight that as shown in Fig. 2.2 the output dimension of the convolution operation is not necessarily the same as that of the input. In the simple case denoted in Eq. (2.3) we have that the output would be $(W - F + 1) \times (H - F + 1) \times K$, which assumes we do not allow any mismatch between the filter indexes and tine input indexes. This is known as *valid* padding. Additionally a convolution may implement what is known as a *stride*, where the layer skips over certain indexes. For a full discussion of padding and stride nuances, we defer interested readers to [1] for a more comprehensive treatment.

Typically, convolutional neural network architectures employ a series of consecutive convolutional layers followed by a series of fully connected layers. We visualize the simplest version of a convolutional neural network in Fig. 2.3.

Recurrent Neural Networks

Recurrent layers are designed to handle sequential data, where the order of data points is important. They are the fundamental building blocks of Recurrent Neural Networks (RNNs), which are used for tasks such as time series prediction, natural language processing, and speech recognition.

In a recurrent layer, the output at each time step depends not only on the current input but also on the hidden state from the previous time step. This allows the network to maintain a memory of past inputs, capturing temporal dependencies.

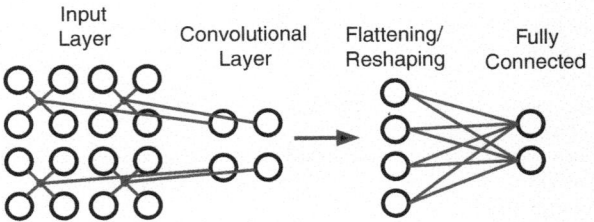

Fig. 2.3 A diagram of a simplistic convolutional neural network. The first layer is depicted as a convolution, followed by a flattening transform, followed by a fully connected layer which would represent the output of the model

Formally, given an input sequence $\mathbf{x} = [\mathbf{x}_1, \mathbf{x}_2, \ldots, \mathbf{x}_T]$, the hidden state $\mathbf{h}_t$ at time step t is computed as:

$$\mathbf{h}_t = \sigma(\mathbf{W}_h \mathbf{h}_{t-1} + \mathbf{W}_x \mathbf{x}_t + \mathbf{b}_h) \tag{2.4}$$

where $\mathbf{W}_h$ and $\mathbf{W}_x$ are weight matrices, $\mathbf{b}_h$ is a bias term, and σ is a non-linear activation function. Variants of the basic RNN, such as Long Short-Term Memory (LSTM) networks and Gated Recurrent Units (GRUs), have been developed to address issues like vanishing gradients and to better capture long-term dependencies.

2.1.2.2 Learning in Neural Networks

In this section, we briefly describe the process of *learning* in neural networks. We first describe the maximum likelihood principle which establishes the learning objective as an optimisation problem. We then describe and visualise stochastic gradient descent as a method to solve the optimisation problem.

Maximum Likelihood and Learning as Optimisation

Stemming from probability and statistics, the principle of maximum likelihood is used to fit the parameters of a model such that they faithfully reproduce the observed data. In our previous sections on classification (Sect. 2.1.1.1) and regression (Sect. 2.1.1.2) we have stated particular likelihoods which correspond to a probabilistic interpretation of the outputs of a machine learning model, such as a neural network.[1] Provided that we represent the loss function abstractly with the *loss* function $\mathcal{L} : \mathbb{R}^{m \times m} \to \mathbb{R}$ which can be seen as the inverse of the likelihood function and maps pairs of outputs (namely a prediction and ground truth value) to a real value, where lower values indicate that the model output and ground-truth are more similar. The task of *learning* the best model parameters is then posed as the

---

[1] We refer interested readers to the following probabilistic treatment of machine learning [2].

2.1 Preliminaries and Technical Background

following optimisation problem:

$$\theta^\star = \arg\min_\theta \mathbb{E}_{(x,y)\sim\mathcal{D}}\left[\mathcal{L}(f^\theta(x), y)\right] \tag{2.5}$$

Given the general non-linearity of the function f^θ solving for $\theta^\star$ is not analytically tractable. Instead, a variety of optimisation methods are employed to approximately find $\theta^\star$. We focus here on gradient descent.

Gradient Descent

Gradient descent is the core algorithm behind the *learning* portion of many machine learning algorithms. The concept behind this important algorithm is simple: if you want to find a/the minimum of a function, pick a random spot on the function and then iteratively take steps in the direction of steepest descent. Mathematically, the direction of steepest descent is the gradient of the function which we will denote with the ∇ operator. This operator will take a subscript to tell us what we are taking the gradient with respect to; for example, ∇_θ tells us to take the gradient of the proceeding function with respect to θ. Given this, we are now ready to state the most basic form of gradient descent:

$$\theta^{(t+1)} = \theta^{(t)} + \alpha \nabla_\theta \mathcal{L}(f^\theta, \mathbf{X}, \mathbf{y}), \tag{2.6}$$

where we assume without loss of generality that $\theta^{(0)} \sim \mathcal{N}(0, 1)$ and that α is constant that we call the *learning rate*. In what follows we will visualise how this iterative equation behaves for different selections of its user-defined parameters. Throughout this we will think of the loss function value, $\mathcal{L}(f^\theta, \mathbf{X}, \mathbf{y})$, as being a *height* or altitude that we would like to minimise. Specifically, visualising the parameter θ as a puck, we will consider pushing the puck in a direction that minimises its height.

Visualising Gradient Descent in Logistic Regression

To visualise how gradient descent works, we will use a simple logistic regression task and visualise the location of the puck and the "height" of the function it is on with red indicated high and blue indicating low as seen in Figs. 2.4 and 2.5. Below we discuss and visualise the working of Eq. (2.6) using different values of α and different terminal values of t.

Consider travelling along a bowl shaped landscape (which as we will see is a reasonable conceptual approximation). The learning rate can be conceptualised as how *fast* you are moving. In the case of Eq. (2.6) the learning rate is also called the *step size*. In Fig. 2.4 subplot b, we can see that there is a noticeable distance between red points indicating a large step size. Indeed the legend indicates that this is the largest learning rate we explored at $\alpha = 0.75$. Though each of our learning "trajectories" (i.e., paths through the loss landscape) start in a different place and therefore have a different starting loss, we can see in Fig. 2.4 subplot a that the

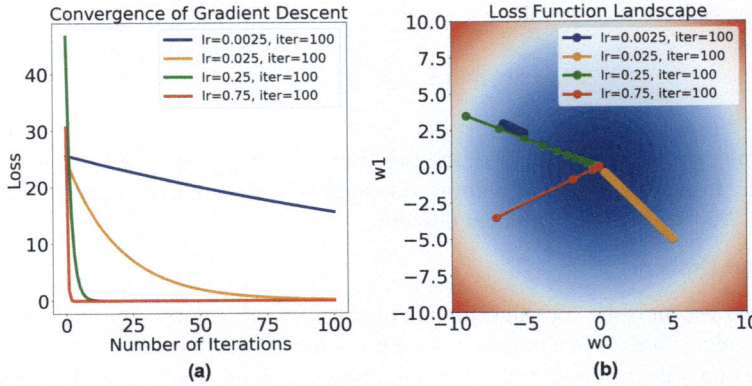

Fig. 2.4 Effect of learning rate on the learning of our logistic regression model. (**a**) The convergence of the loss, we can interpret this as the *height* of the puck in the bowl. (**b**) A visualisation of the path the parameter (puck) as it moves in the bowl

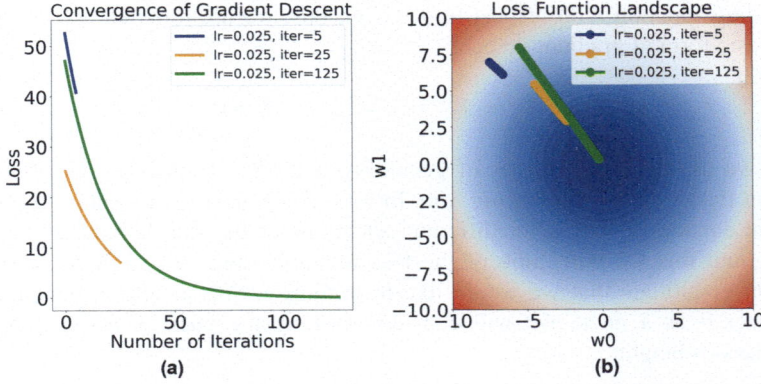

Fig. 2.5 Effect of number of iterations on the learning of our logistic regression model. (**a**) The convergence of the loss, we can interpret this as the *height* of the puck in the bowl. (**b**) A visualisation of the path the parameter (puck) as it moves in the bowl

primary effect of the learning rate in this case is to speed up the rate at which we learn.

Where we introduced the intuitive interpretation of the learning rate as the speed of our puck the number of iterations can be considered the amount of time we allow the puck to travel. We can see in Fig. 2.5 that the longer we allow the puck to travel, the closer it gets to our optimal value.

2.1.3 Basics of Explainability

Machine learning models, and deep ones in particular, are known to excel in tasks that require the ability to process vast amount of data. For instance, it is well-known that neural networks can solve classification problems over images with unparalleled accuracy. However, many applications require more than just an accurate prediction. This is typically the case when machine learning is deployed in safety-critical or consequential decision-making tasks. For example, consider a setting in which a neural network is used to predict the creditworthiness of a person applying for a loan. Irrespective of the decision produced, understanding why such a decision was reached is extremely important. This is also advocated in several frameworks for AI ethics and in regulations. For example, the Organisation for Economic Co-operation and Development states that information should be provided "to enable those adversely affected by an AI system to challenge its outcome based on plain and easy-to-understand information on the factors, and the logic that served as the basis for the prediction, recommendation or decision" (Principle[2] 1.3).

Explanations methods can help improve the interpretability of machine learning models by shedding light on their decision-making mechanisms. This is typically achieved by relating feature values of a given input instance to the corresponding prediction made by the model. The details of how such relation is identified may vary considerably between existing solutions. However, explanation algorithms can be classified based on some key properties discussed below.

The first major distinction concerns whether explainability is achieved *ante-hoc* or *post-hoc*. Ante-hoc explainability refers to settings where the architectural complexity of a machine learning model is reduced considerably during training, thus resulting in models that are interpretable by construction. Typical examples include (small) decision trees or (sparse) linear models. Post-hoc explainability instead refers to a setting where the complexity of the model is unaltered, and explanations are generated by external algorithms applied after training.

Another important difference is between *local* explainability methods and *global* ones. The former explain the output of a machine learning model with respect to a given input instance. For example, when explaining why a loan application was declined, a local method might related that decision to specific characteristics of the person applying for it. On the contrary, a global explanation method tries to give a window on how a machine learning model makes decision without focusing on specific inputs, thus providing a more holistic view on its behaviour.

Explanation algorithms can also be separated into those operating in a *model-agnostic* way and those that rely on *model-specific* properties of the machine learning model at hand. As a result, the former generate explanations that only look at input-output behaviours of a model, whereas the latter require access to

---

[2] https://oecd.ai/en/dashboards/ai-principles/P7.

its internals (such as weights in a neural network). Moreover, agnostic approaches are typically not tied to specific data types and therefore afford more flexibility.

2.2 The Wider World of Explainability

In this section, we briefly describe some of the seminal works in the explainability literature. While many of these works will be discussed in further detail in later chapters (we reference when this is the case), in this section we aim at providing readers with a general sense of the explainability literature without making an attempt to be comprehensive. The interested reader may refer to recent surveys [3] for further details on the explainability algorithms.

The first popular class of explanation methods we consider is *feature attribution*. These methods aim to explain the prediction of a machine learning model $\hat{y}$ in terms of a score $S(x)$ for each component of the corresponding input vector x. These scores are then used to indicate how much each feature contributed to the predicted classification $\hat{y}$, either negatively or positively. Several attribution approaches exist, including perturbation-based methods such as LIME [4] or SHAP [5], and gradient-based such as saliency maps [6]. The latter specialise feature attribution to image data in the context of neural networks. In particular, saliency-based approaches assign a score to each pixel in the input image according to its importance towards defining the predicted classification. Saliency-based explanations and robustness thereof will be introduced formally in Chap. 4

Differently from feature attribution methods, *counterfactual explanations* aim to describe how an input should have been different for a different prediction to be computed [7]. In the context of deep learning, a counterfactual explanation is typically characterised in terms of minimal changes to x that would result in an x' such that $f^\theta(x) \neq f^\theta(x')$. Counterfactual explanations are often used to provide recourse recommendations to users that have been negatively affected by the predictions of a machine learning model. When used to this end, counterfactuals are defined in terms of changes aimed at attaining specific class scores, i.e. $f^\theta(x) \neq f^\theta(x')$ with $f^\theta(x') = j$, $j \in [c]$. Counterfactual explanations and robustness thereof will be introduced formally in Chap. 3.

Another popular explanation strategy consists in identifying *prototypes* [8], i.e. inputs that summarise the behaviour of a machine learning model across several, similar instances. Prototypes are useful as they represent examples which a user can use to understand the reasoning of the machine learning model by, e.g. comparing the predictions that a model produces for the original input and its corresponding prototypes. Prototypical explanations for an input x can be computed in several ways [3]. For example, one could return another input x', similar to x, chosen from the training dataset $\mathcal{D}$. Similarly, a prototype could be obtained by first identifying the cluster of inputs from $\mathcal{D}$ which x belongs to, and then returning the centroid of said cluster.

Finally, *rule-based* explanation approaches aim to explain the behaviour of a model by means of simple, logic-based rules extracted from it. Local rule-based explainers such as ANCHOR [9] and LORE [10] produce logical rules that capture the decision making of a model around a given input **x**. Methods such as GLocalX [11] can generate global rules by iteratively merging local rules and pruning those that are logically subsumed by aggregated rules.

References

1. Ian Goodfellow, Yoshua Bengio, and Aaron Courville. *Deep Learning*. MIT Press, 2016. http://www.deeplearningbook.org.
2. Kevin P Murphy. *Probabilistic machine learning: an introduction*. MIT press, 2022.
3. Francesco Bodria, Fosca Giannotti, Riccardo Guidotti, Francesca Naretto, Dino Pedreschi, and Salvatore Rinzivillo. Benchmarking and survey of explanation methods for black box models. *Data Min. Knowl. Discov.*, 37(5):1719–1778, 2023.
4. Marco Túlio Ribeiro, Sameer Singh, and Carlos Guestrin. "why should I trust you?": Explaining the predictions of any classifier. In Balaji Krishnapuram, Mohak Shah, Alexander J. Smola, Charu C. Aggarwal, Dou Shen, and Rajeev Rastogi, editors, *Proceedings of the 22nd ACM SIGKDD International Conference on Knowledge Discovery and Data Mining, San Francisco, CA, USA, August 13-17, 2016*, pages 1135–1144. ACM, 2016.
5. Scott M. Lundberg and Su-In Lee. A unified approach to interpreting model predictions. In Isabelle Guyon, Ulrike von Luxburg, Samy Bengio, Hanna M. Wallach, Rob Fergus, S. V. N. Vishwanathan, and Roman Garnett, editors, *Advances in Neural Information Processing Systems 30: Annual Conference on Neural Information Processing Systems 2017, December 4-9, 2017, Long Beach, CA, USA*, pages 4765–4774, 2017.
6. Mukund Sundararajan, Ankur Taly, and Qiqi Yan. Axiomatic attribution for deep networks. In Doina Precup and Yee Whye Teh, editors, *Proceedings of the 34th International Conference on Machine Learning, ICML 2017, Sydney, NSW, Australia, 6–11 August 2017*, volume 70 of *Proceedings of Machine Learning Research*, pages 3319–3328. PMLR, 2017.
7. Sandra Wachter, Brent D. Mittelstadt, and Chris Russell. Counterfactual explanations without opening the black box: Automated decisions and the GDPR. *Harv. JL & Tech.*, 31:841, 2017.
8. Karthik S. Gurumoorthy, Amit Dhurandhar, Guillermo A. Cecchi, and Charu C. Aggarwal. Efficient data representation by selecting prototypes with importance weights. In Jianyong Wang, Kyuseok Shim, and Xindong Wu, editors, *2019 IEEE International Conference on Data Mining, ICDM 2019, Beijing, China, November 8–11, 2019*, pages 260–269. IEEE, 2019.
9. Marco Túlio Ribeiro, Sameer Singh, and Carlos Guestrin. Anchors: High-precision model-agnostic explanations. In Sheila A. McIlraith and Kilian Q. Weinberger, editors, *Proceedings of the Thirty-Second AAAI Conference on Artificial Intelligence, (AAAI-18), the 30th innovative Applications of Artificial Intelligence (IAAI-18), and the 8th AAAI Symposium on Educational Advances in Artificial Intelligence (EAAI-18), New Orleans, Louisiana, USA, February 2–7, 2018*, pages 1527–1535. AAAI Press, 2018.
10. Riccardo Guidotti, Anna Monreale, Fosca Giannotti, Dino Pedreschi, Salvatore Ruggieri, and Franco Turini. Factual and counterfactual explanations for black box decision making. *IEEE Intell. Syst.*, 34(6):14–23, 2019.
11. Mattia Setzu, Riccardo Guidotti, Anna Monreale, Franco Turini, Dino Pedreschi, and Fosca Giannotti. Glocalx - from local to global explanations of black box AI models. *Artif. Intell.*, 294:103457, 2021.

Chapter 3
Robustness of Counterfactual Explanations

3.1 Introduction

Counterfactual explanations can help clarify the decisions of a (black-box) machine learning model by showing how an input should be altered to obtain a different decision from it. This is particularly useful in the context of algorithmic recourse, where the aim is to provide recourse recommendations to individuals that have been negatively affected by the decisions of a model. Counterfactual explanations have been advocated as being particularly suited for this, due to their appeal to users [1], intelligibility [2] and alignment with human reasoning [3].

When dealing with recourse in the context of consequential decision-making tasks, e.g. in financial or medical settings, it is of utmost importance that the recourse recommendations provided are valid and yield the intended change in outcome. However, recent work has demonstrated that state-of-the-art algorithms for generating counterfactual explanations suffer from major drawbacks pertaining to the *robustness* of the explanations they generate. In particular, it has been shown that recourse recommendations provided by many counterfactual explanation algorithms are extremely susceptible to small changes occurring in the setting they were generated for, and may thus fail to achieve their intended outcome. Clearly such brittleness is undesirable, as a lack of robustness may lead to users questioning whether the explanations are actually capturing the decision making process of the model under analysis, ultimately jeopardising the explanatory function of counterfactuals.

Several approaches have been proposed to mitigate risks and concerns arising from a lack of robustness in counterfactual explanations. This chapter offers a unified presentation of these approaches. In particular, we will first introduce basic notions on counterfactuals explanations in Sect. 3.2. We will then proceed with a discussion on different notions of robustness that have been analysed in the literature in Sect. 3.3, covering *robustness to input changes* in Sect. 3.3.1, *model changes* in Sect. 3.3.2, *model multiplicity* in Sect. 3.3.3 and *noisy execution* in Sect. 3.3.4. For each notion, we will overview algorithms devised to solve the robustness problem and metrics to evaluate the robustness of counterfactual explanations. We will then conclude this chapter with Sect. 3.4, which provides a discussion on open problems and challenges that characterise this active field of research.

3.2 Counterfactual Explanations

For ease of exposition, we will focus on machine learning models trained to solve binary classification tasks, i.e. where only two classes exist. However, many of the results discussed in this chapter generalise to the more general setting of multi-class classification and regression. Assume a classification model $f^\theta : \mathbb{R}^{n_x} \to \{0, 1\}$ where $\mathbb{R}^{n_x}$ is the input space and $\{0, 1\}$ is a set of output labels. Given an input $\mathbf{x} \in \mathbb{R}^{n_x}$ and a model f^θ, mainstream approaches characterise counterfactual explanations in terms of the solutions of the following optimisation problem [4]:

$$\arg\min_{\mathbf{x}' \in \mathbb{R}^{n_x}} cost(\mathbf{x}, \mathbf{x}') \text{ s.t. } f^\theta(\mathbf{x}') \neq f^\theta(\mathbf{x}) \tag{3.1}$$

where $cost : \mathbb{R}^{n_x} \times \mathbb{R}^{n_x} \to \mathbb{R}^+$ is a suitable cost metric defined over the input space.

Following this formulation, a *valid* counterfactual explanation can be defined as the closest input $\mathbf{x}'$ to $\mathbf{x}$ that changes the classification produced by f^θ. This change is typically assumed to move from a negative classification outcome to a positive one, although different formalisations are possible depending on the task at hand. As regards the cost metric, typical examples include, e.g. the ℓ_1 or ℓ_∞ norms. Most approaches require minimising one such cost metric to account for the real-world effort that would be required for a user to move from the scenario captured by $\mathbf{x}$ to a new one described by $\mathbf{x}'$, as discussed in Example 3.1.

Example 3.1 Consider a (fictional) loan application where Bob applies for a loan with a bank which uses a machine learning model to process the application and decide whether the loan should be granted or not. For illustration, assume Bob's loan application is modelled by an input $\mathbf{x}$ with features 25 years of *age*, $15, 000 *loan amount* and $20, 000 *salary*.

Deciding whether Bob should be granted a loan can be cast as a binary classification task, where the machine learning model is trained to predict one of two labels, corresponding to whether Bob will be able to repay the loan (e.g. label 1) or not (e.g. label 0) and thus whether the loan application should be accepted or not.

Assume that Bob's application is initially rejected based on the prediction made by the classifier that Bob will not be able to repay the loan back, as shown in Fig. 3.1. A possible counterfactual explanation for the rejection could be an altered input $\mathbf{x}'$, where a salary of $25, 000 (with the other features unchanged) would result in the loan being accepted, thus giving Bob an idea of what needs to be

3.2 Counterfactual Explanations

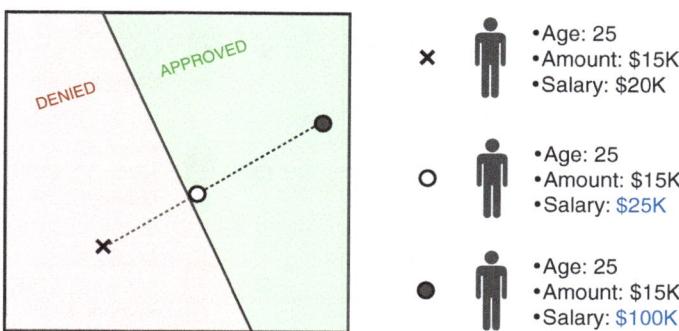

Fig. 3.1 A 2-D visualisation of the input space of a (fictional) binary classifier, which partitions the space into two classes (denied, approved). Two counterfactual explanations are generated for the original input (*times*): one that minimises cost (*open circle*) and one that does not (*filled circle*). Minimising distance affects the magnitude of changes prescribed by each counterfactual

changes in his application. This example also clarifies the role played by the cost metric in Eq. (3.1). Imagine the counterfactual had suggested increasing the salary to $100, 000, disregarding any notion of minimality. Bob might have troubles implementing this change, thus rendering the counterfactual not useful in practice.

Computing an exact solution for the problem presented in Eq. (3.1) may be viable only for certain types of machine learning models. For instance, deep neural networks using piece-wise linear activation functions are known to be encodable into Mixed-Integer Linear Programs [5], thus enabling exact approaches to counterfactual generation [6]. For more general classes of differentiable models, a relaxed formulation is typically considered instead:

$$\underset{\mathbf{x}' \in \mathbb{R}^{n_x}}{\arg\min} \; loss(f^\theta(\mathbf{x}'), f^\theta(\mathbf{x})) + \lambda \cdot cost(x, x') \qquad (3.2)$$

where *loss* is a differentiable loss function that guides the search towards a valid counterfactual explanation $\mathbf{x}'$, and λ is a parameter dictating a trade-off between validity and cost.

The basic formulation presented in Eqs. (3.1) and (3.2) has been extended to account for other important properties. For instance, counterfactual explanations are typically required to be *actionable* [7] and only suggest changes to mutable features that can be realistically modified by the user (see Fig. 3.2 for an illustration). This is particularly important when counterfactuals are used to provide recourse: suggesting to change, e.g. one's *age*, would result in a recommendation that is not feasible in practice. On the contrary, a change in salary would be considered actionable and thus practically viable. Another important requirement is *plausibility* [8], which requires that counterfactual explanations try to adhere as much as possible to the data manifold, to avoid suggesting unrealistic changes (see Fig. 3.3 for an illustration). *Sparsity* [4] is also deemed important, whereby counterfactual

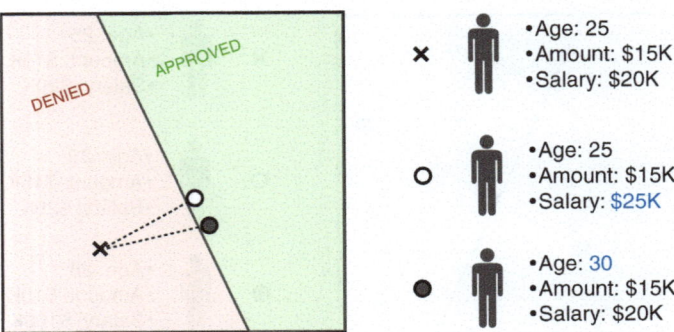

Fig. 3.2 An actionable counterfactual explanation (*open circle*) only suggests modifications to mutable features, such as increasing salary to $25,000$. On the contrary, a non actionable explanation (*filled circle*) may suggest to increase age to 30, which is clearly impossible for the user (*times*)

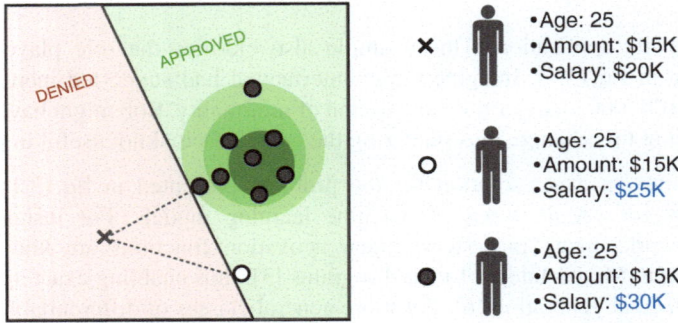

Fig. 3.3 A plausible counterfactual explanation (*filled circle*) lies close to the data manifold (*shaded green areas*) and is likely to be more informative. Instead, a counterfactual explanation lacking plausibility (*open circle*) may lie in regions of the input space with low sample density. In this example, we notice that prioritising plausibility may incur in increased recourse cost

explanations suggesting isolated changes on few features are to be preferred over explanations that change a large number of features. This is crucial to ensure that the counterfactual does not overload the user with too much information which would only hinder the explanation process. Finally, *diversity* [9] advocates the generation of a *set* of (diverse) counterfactual explanations for each input as opposed to returning a single one. This is done to provide a better approximation of local decision boundaries of machine learning models, thus strengthening the explanatory power of counterfactuals (see Fig. 3.4 for an illustration). It is important to stress that these properties might conflict each other, e.g. generating a plausible counterfactual explanation might increase its cost. We refer to [10] for more details on counterfactual explanations and their commonly studied properties.

3.3 Robustness of Counterfactual Explanations

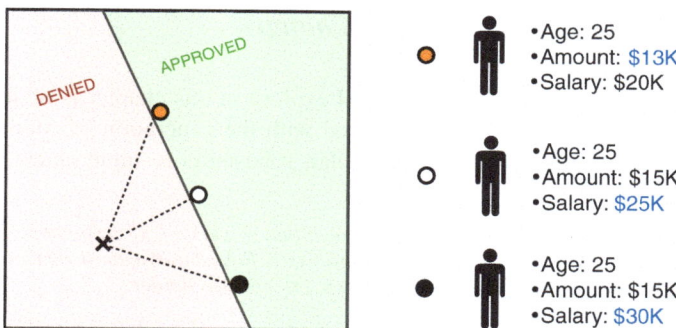

Fig. 3.4 A set of diverse counterfactual explanations (●, ○, ◐) is returned to the user and can be used to better approximate the decision boundary of the machine learning model in the vicinity of the original input (*times*)

3.3 Robustness of Counterfactual Explanations

As argued in the introduction of this book, robustness plays a crucial role in explainability. This becomes evident when counterfactual explanations are used to provide recourse recommendations in consequential decision-making tasks, as a lack of robustness in these settings might have considerable impacts on different stakeholders. Indeed, users might end up receiving brittle recourse recommendations that may ultimately lead to unfavourable outcomes, leaving those who proposed such recourse options liable. Worryingly, this problem is not just mere speculation. As we will see, state-of-the-art explanation algorithms have been shown to be prone to generating counterfactual explanations whose validity may be easily compromised, which raises questions about their reliability. A recent survey on this topic [11] identified four main sources of fragility in counterfactual explanations, namely:

1. *input changes*,
2. *model changes*,
3. *model multiplicity* and
4. *noisy execution*.

In the rest of this chapter we will analyse these four scenarios, focusing on the different ways in which robustness has been formalised. We will then discuss the implications that a lack of robustness may have in each of them, together with solutions proposed to generate counterfactual explanations with improved robustness guarantees.

3.3.1 Robustness Against Input Changes

The first notion of robustness that we will explore in this chapter is related to the behaviour of explanation algorithms tasked with the generation of counterfactual explanations for similar inputs. In particular, robustness to input changes can be abstractly summarised as follows.

> Assume two inputs $\mathbf{x}_1, \mathbf{x}_2$ and a model f^θ such that $f^\theta(\mathbf{x}_1) = f^\theta(\mathbf{x}_2)$. Let $\mathbf{x}'_1, \mathbf{x}'_2$ be counterfactual explanations for $\mathbf{x}_1, \mathbf{x}_2$, respectively. Robustness against input changes requires that whenever $\mathbf{x}_1, \mathbf{x}_2$ are **similar**, then $\mathbf{x}'_1, \mathbf{x}'_2$ are also **similar**.

Intuitively, one would expect that slight perturbations to the input of a machine learning model would not result in qualitatively different counterfactual explanations. This is particularly important in the context of algorithmic recourse, where one would want to ensure that similar individuals receive similar recourse recommendations. To see why this is important let us consider Example 3.2.

Example 3.2 Consider two customers, Alice and Bob, both applying for a loan at the same bank. Bob is 25 years old, has an annual income of $20,000 and applies for a $15,000 loan. Alice is 24 years old, has the same annual income as Bob and requests the same loan amount. Both applications are initially rejected by the bank, as both Alice and Bob are predicted not to be able to repay their loans in time by the bank's machine learning model, as depicted in Fig. 3.5. Now assume an explanation algorithm is used by the bank to provide counterfactual explanations to both users. Bob is told that a salary increase of $5000 would be sufficient for the loan to be granted, whereas Alice is asked to increase her salary by $20,000. Upon receiving such different recommendations, Alice may start wondering why such an unfavourable recourse option was given to her. After all, she earns the same as Bob!

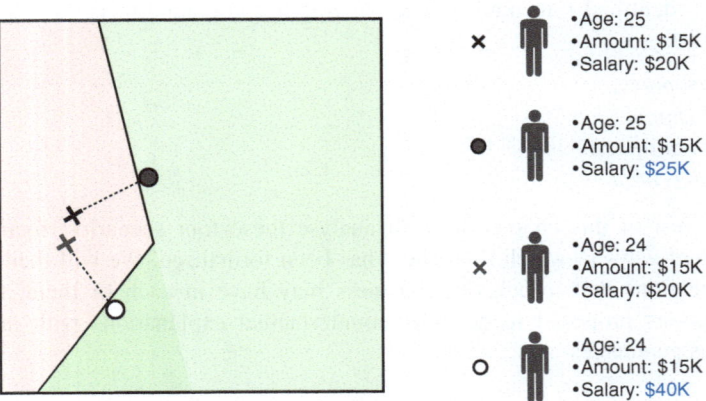

Fig. 3.5 An explanation algorithm lacking robustness to input changes may return very different counterfactual explanations (*filled circle, open circle*) for similar users (*black times, gray times*)

3.3 Robustness of Counterfactual Explanations

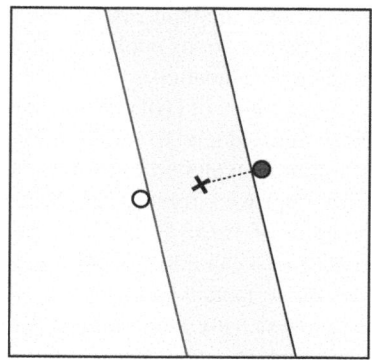

Fig. 3.6 When multiple counterfactuals may exist (*filled circle, open circle*) for an input (*times*), algorithms returning a single (nearest) counterfactual are bound to lack robustness

The example reveals (at least) two potential implications that a lack of robustness to input changes may have. First, it may raise doubts about the extent to which counterfactual explanations are reflective of the logic underpinning the machine learning model. In our example both Alice and Bob earn the same and applied for the same loan amount, why would the machine learning model require such a difference between salaries for the loans to be granted? This leads to another important problem: moral hazard. As noted in [12] an explanation algorithm that lacks robustness may lead to unfair recommendations that might damage individuals belonging to disadvantaged groups. To complicate things further, recent work [13] has shown that deep learning models can be trained to exhibit this kind of behaviour. In particular, the authors have shown that neural networks can be trained in such a way that individuals belonging to a non-protected group can always obtain a lower cost recourse when compared to protected groups, raising fairness concerns.

There are many possible reasons why a counterfactual explanation algorithm may lack robustness to input changes. Firstly, as noted in [13], hill-climbing methods that are typically used to solve Eq. (3.2) are known to be susceptible to input perturbations and may thus lack robustness. Secondly, counterfactual explanations for a given input may not be unique [14] as exemplified in Fig. 3.6. When this is the case, explanation algorithms that operate by returning the nearest counterfactual explanation (cf. Eqs. (3.1) and (3.2)) may exhibit unstable behaviours. In this case, returning a single explanation may fail to provide a good characterisation of the decision boundary of a model and may thus lead to oscillatory behaviours between potential counterfactuals whenever the input to be explained changes slightly.

This problem was first studied in [15], where the authors present a theoretical analysis of the behaviour of counterfactual explanation algorithms applied to binary classification models. The authors propose to quantify robustness of counterfactual explanations in terms of their *local instability*. In particular, given a classification model f^θ, an input $\mathbf{x}_1$ and a counterfactual $\mathbf{x}'_1$ such that $f^\theta(\mathbf{x}_1) \neq f^\theta(\mathbf{x}'_1)$, local instability is defined as the expected distance between $\mathbf{x}'_1$ and the counterfactual explanations computed for inputs similar to $\mathbf{x}$, i.e. $\mathbb{E}_{\mathbf{x}_2 \sim S(\mathbf{x}_1)}[d(\mathbf{x}'_1, \mathbf{x}'_2)]$, where $S(\mathbf{x}_1)$ denotes a set of inputs that are similar to $\mathbf{x}_1$ and d is a distance metric defined over the input space. Intuitively, local instability measures the influence that small perturbations applied to the inputs of explanation algorithms can have on the resulting explanation.

The authors of [16] observe that counterfactual explanations that lie deeper in the data manifold tend to exhibit higher levels of robustness to input perturbations. This intuition is exploited to propose a method to generate counterfactual explanations with improved robustness. In a nutshell, the authors modify the optimisation problem of Eq. (3.2) and add an additional loss term that accounts for the *local density* of a candidate solution. This is computed as a normalised distance between the counterfactual and its neighbouring instances. Intuitively, a counterfactual lying in a low-density area is likely to lack robustness and thus should not be chosen. Therefore, by accounting for this measure during counterfactual search, the authors provide a procedure to only return counterfactual explanations that fit the data manifold to a higher extent.

Another solution to the robustness problem is presented in [17], where the authors propose to relax the proximity requirement for counterfactuals and use sets of (diverse) explanations to improve the stability of explanation algorithms. In particular, the authors define two types of counterfactual explanations: *strong counterfactuals*, which lie exactly on the decision boundary, and ε-*approximate* ones, whose distance from the decision boundary is at most ε. Then, given two sets of counterfactual explanations $E(\mathbf{x}_1)$, $E(\mathbf{x}_2)$ computed for similar inputs $\mathbf{x}_1, \mathbf{x}_2$, the authors introduce a notion of distance between two sets of counterfactual explanations as:

$$\text{set-distance}_{\Sigma}^{d}(E(\mathbf{x}_1), E(\mathbf{x}_2)) = \frac{1}{2 \cdot |E(\mathbf{x}_1)|} \sum_{\mathbf{x}'_1 \in E(\mathbf{x}_1)} \min_{\mathbf{x}'_2 \in E(\mathbf{x}_2)} d(\mathbf{x}'_1, \mathbf{x}'_2)$$
$$+ \frac{1}{2 \cdot |E(\mathbf{x}_2)|} \sum_{\mathbf{x}'_2 \in E(\mathbf{x}_2)} \min_{\mathbf{x}'_1 \in E(\mathbf{x}_1)} d(\mathbf{x}'_2, \mathbf{x}'_1).$$

Intuitively, low values of set-distance$_{\Sigma}^{d}$ correspond to scenarios in which for each counterfactual explanation in $E(\mathbf{x}_1)$ there exist a close one in $E(\mathbf{x}_2)$. Using this metric, the authors propose a formal notion of robustness, termed *weak ϵ-robustness*, which requires that strong counterfactuals computed for an input $\mathbf{x}_1$ are retained when moving towards $\mathbf{x}_2$ by at most ϵ and thus become ε-approximate counterfactuals for $\mathbf{x}_2$ (and vice versa). An approximate algorithm is also derived using nearest neighbour counterfactual explanation algorithms [18] to compute counterfactual explanations with improved robustness guarantees.

To conclude this section, we would like to draw attention to other problems from the literature on counterfactual explanations that share interesting connections with robustness to input changes. In particular, we see the problem of computing counterfactuals explanations for inputs with missing feature values [19] as particularly interesting. In this setting, one is typically interested in computing explanations that are valid for any missing feature value, thus addressing a special form of robustness to input changes. Also related is the problem of computing group counterfactuals [20–22], i.e. counterfactual explanations that produce recourse recommendations that are valid for sets of input instances. We believe it would be interesting to study these different scenarios under the unifying perspective of robustness to input changes, to understand the advantages and limitations of each approach and enable potential cross-fertilisation.

3.3.2 Robustness Against Model Changes

Model changes are perhaps the most commonly studied scenario in which the robustness of a counterfactual may be affected [11]. These changes are typically defined as alterations in the parameters of a machine learning model that do not modify its architecture, e.g. as resulting from fine-tuning on inputs drawn from a data distribution that shift over time. Irrespective of how model changes happen, they can have serious implications in the context of counterfactual explanations and recourse, as shown in Example 3.3.

Example 3.3 (Continuing from Example 3.1) After being rejected, Bob would expect that when applying again, achieving a salary increase of $5000 would result in the application being successful, as captured by the original counterfactual explanation provided by the bank. However, imagine what would happen if a fine-tuning step occurs while Bob is working toward achieving a higher salary. If the original counterfactual provided to Bob lacks robustness, the validity of the recourse recommendation suggesting to increase salary may still result in a rejected application, leaving the bank liable due to their conflicting statements.

When presented with scenarios such as the one described in Example 3.3, two possible strategies could be conceived to deal with model changes. The first would be to further update the machine learning model to guarantee that customers that were offered recourse before the model update will still obtain a favourable outcome when applying again. However, this might result in a considerable financial loss for the bank, which may be required to offer loans to customers that are not predicted to be creditworthy by the retrained model. Another option for the bank would be to update the model while also including constraints that enforce the validity of previously generated recourse recommendations. However, this option may result in models that do not reflect the shifted data distribution and are thus sub-optimal (Fig. 3.7).

To overcome the challenges above, new approaches have been proposed to ensure robustness of counterfactual explanations under model shifts, as abstractly summarised in the following.

Assume an input $\mathbf{x}$ and a model f^θ. Let $\mathbf{x}'$ be a counterfactual explanation for $\mathbf{x}$. Robustness against model changes requires that whenever the model f^θ **changes** to $f^{\theta'}$, and this change is **sufficiently small**, then $f^\theta(\mathbf{x}') = f^{\theta'}(\mathbf{x}')$.

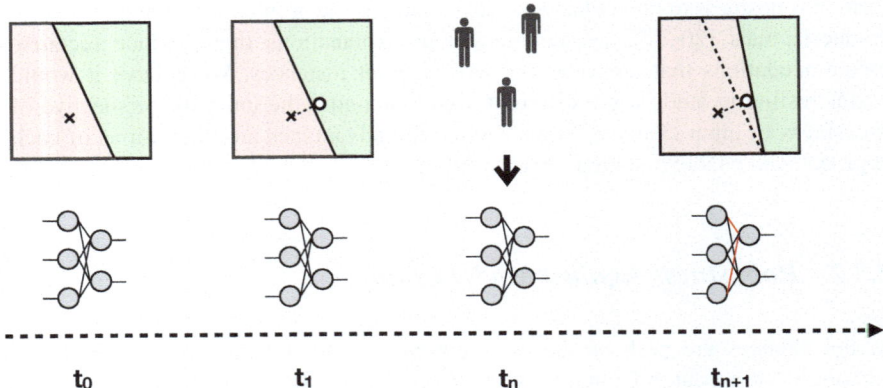

Fig. 3.7 Illustration of how a lack of robustness to model changes may affect the validity of a counterfactual explanation. In this example, a neural network initially classifies an input (*times*) as not creditworthy and a counterfactual explanation is computed (*open circle*). Later, new data become available and the bank decides to fine-tune its existing model to incorporate this new information. As a result of this step, some parameters in the network may be updated, causing the decision boundary of the model to slightly change. Ultimately, this results in the old counterfactual explanation not being valid anymore

Different notions of what can be considered a sufficiently small change have been proposed in the literature. For example, [23] first considered *plausible model changes*, which are defined as updates to the parameters $\theta \in \mathbb{R}^n$ of a model f^θ whose magnitude is bounded by a small constant. Formally, plausible model shifts are captured by:

$$\Delta = \{\delta \in \mathbb{R}^n \mid \|\delta\|_p \leq \varepsilon\}$$

for a fixed constant $\varepsilon \in \mathbb{R}$ and a p-norm $\|\cdot\|_p$ defined as usual. Intuitively, plausible model shifts capture small changes to the parameters of the original model f^θ that are assumed not to alter its overall behaviour.

A different notion of model changes is offered in [24], where *naturally-occurring model changes* are introduced for deep learning classifiers. A set of model changes Δ is said to be naturally occurring if for a (randomly) chosen model change $\delta \in \Delta$ and $f^{\theta'} = f^{\theta+\delta}$ being the new classifier obtained after applying δ to θ, the following conditions hold [24]:

1. $\mathbb{E}[f^{\theta'}(\mathbf{x})] = f^\theta(\mathbf{x})$; where the expectation is over the randomness of $f^{\theta'}$ given a fixed input $\mathbf{x}$;
2. $\mathrm{Var}[f^{\theta'}(\mathbf{x})] = v_\mathbf{x}$, where $v_\mathbf{x}$ represents the maximum variance of the prediction of $f^{\theta'}(\mathbf{x})$, and whenever $\mathbf{x}$ lies on the data manifold, $v_\mathbf{x}$ is upper bounded by a small constant v;
3. If f^θ is Lipschitz continuous for some $\gamma_1 \in \mathbb{R}$, then $f^{\theta'}$ is also Lipschitz continuous for some $\gamma_2 \in \mathbb{R}$.

3.3 Robustness of Counterfactual Explanations

Following the definition above, a naturally-occurring model change allows the application of arbitrary changes to the parameters of a model as long as its overall behaviour is not affected.

These two notions of model changes capture profoundly different scenarios as demonstrated in [25], and reflect different goals towards solving the robustness problem. Plausible model changes are particularly useful in certification settings, where one may be interested in providing strong guarantees that validity of a counterfactual explanation will not be compromised by changes of (a given) bounded magnitude. On the contrary, naturally-occurring model changes are more suited to describe model alterations that follow natural training dynamics. Depending on the requirements of the application for which counterfactuals are generated, one notion may be more suitable than the other.

Irrespective of how model changes are characterised, metrics are needed to precisely quantify the extent to which a counterfactual explanation can be deemed robust. One common metric is *validity after retraining*, which measures the percentage of counterfactual explanations that remain valid after the model for which they were generated is updated. This metric applies to a wide spectrum of update strategies, including retraining from scratch using shifted datasets [23], retraining on different portions of a dataset [24], incremental retraining on different portions of the same dataset [26] or retraining under changing initial conditions [27]. Despite its versatility, it should be noted that this metric may fail to provide a precise characterisation of robustness in general, as evaluating validity over different sets of counterfactual explanations may result in different outcomes.

This notion can be further specialised depending on the definition of Δ. For example, *recourse action instability* [28] specialises validity after retraining for model changes arising from data deletion queries. These correspond to situations where a model needs to be updated because one training input is removed from the dataset due to, e.g. a user withdrawing their consent to their data being used for training a model as prescribed by data protection laws such as the GDPR.[1]

Counterfactual stability [24, 29] has also been proposed to quantify robustness of a counterfactual $\mathbf{x}'$ under naturally-occurring model changes for a model f^θ as:

$$R_{K,\sigma^2}(\mathbf{x}', f^\theta) = \frac{1}{K} \sum_{\mathbf{x}'' \in N_{\mathbf{x}'}} f^\theta(\mathbf{x}') - \sqrt{\frac{1}{K} \sum_{\mathbf{x}'' \in N_{\mathbf{x}'}} \left(f^\theta(\mathbf{x}') - \frac{1}{K} \sum_{\mathbf{x}'' \in N_{\mathbf{x}'}} \right)^2}$$

where $N_{\mathbf{x}'}$ is a set of K inputs sampled from a Gaussian distribution centred in $\mathbf{x}$ with variance σ^2. Intuitively, a counterfactual explanation $\mathbf{x}$ is stable if the model output is high for $\mathbf{x}$ with low variability in its neighbourhood. This notions was later extended to deep neural networks in [24].

---

[1] https://gdpr-text.com/read/article-17/.

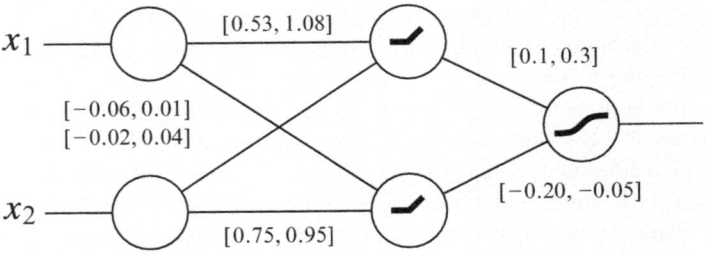

Fig. 3.8 Example of an interval neural network

Having discussed different notions of model changes and related metrics, we are now ready to discuss some algorithmic solutions that have been devised to generate robust counterfactual explanations.

Considering plausible model changes, [23] proposed to generate counterfactual explanations by solving a variant of Eq. (3.2) defined as follows:

$$\arg\min_{\mathbf{x}' \in \mathbb{R}^{n_x}} \max_{\delta \in \Delta} \; loss(f^{\theta+\delta}(\mathbf{x}'), f^{\theta}(\mathbf{x})) + \lambda \cdot cost(x, x')$$

where $loss$ and $cost$ are defined as per Eq. (3.2). By solving the min-max problem above, the authors generate the best counterfactual explanation $\mathbf{x}'$ in terms of $cost$ under the largest model change δ that is allowed by Δ. The problem is solved by means of an iterative procedure where at each step an inner optimisation routine finds the weight perturbation vector δ that maximises the prediction loss. Then, an outer optimisation loop updates the candidate counterfactual explanation by searching for an input optimises the overall cost under the previously computed worst-case model change. A similar idea is elaborated in [25], where the maximisation loop is replaced by a sampling phase which draws perturbation vectors from Δ to derive probabilistic bounds on the validity of a candidate counterfactual explanation.

A worst-case certification approach is presented in [26, 30], where plausible model changes for neural networks are represented by means of Interval Neural Networks (INNs). Intuitively, an INN generalises a standard neural network by considering interval-valued weights, as exemplified in Fig. 3.8. By setting interval weights in accordance with the definition of Δ, INNs provide a compact representation of the family of neural networks that can be obtained from f^{θ} under model changes contained in Δ. This representation allows to reason about validity under model changes in one shot, replacing the maximisation loop of [23] and the sampling of [25]. The authors propose an approach based on Mixed-Integer Linear Programming (MILP) to compute validity certificates for counterfactuals. Leveraging these certificates, an incremental procedure is also proposed in [26], whereby standard (non-robust) algorithms for the generation of counterfactual explanations are combined with an INN certification step to obtain provably robust explanations. This idea was further refined in [31], where the MILP approach is extended to account for the plausiblity of generated counterfactuals. It should be noted that checking the robustness of a counterfactual explanation against plausible model shifts is an NP-complete problem [25]. As such, the scalability of exact algorithms is expected to decrease as the size of the underlying machine learning model increases.

A possible solution to by-pass complexity considerations is to resort to training methods to robustify the explanation generation process. For instance, augmentation techniques leveraging previously generated counterfactuals are proposed in [32]. This approach has been shown to increase the validity of counterfactuals after retraining; however, as already discussed at the beginning of this section, these approaches might result in models that may fail to capture the shifted data distribution faithfully.

Other works [33, 34] compute robust explanations by steering the search for counterfactuals towards points in the input space for which the model f^θ produces outputs with higher scores. Complementing these results, in [27] it is shown that for complex non-linear models, higher class scores alone may not be sufficient to ensure robustness. Therefore, the authors require that the counterfactual explanations be located in regions of the input space of the machine learning model with a low Lipschitz constant, and propose a method that leverages this result during search.

Generate-and-test procedures [24, 29] have been proposed to compute robust counterfactual explanations under naturally-occurring model changes. In a nutshell, these procedure generate a candidate counterfactual explanation by using standard (non-robust) methods and assess its stability as defined in [29]. When the counterfactual passes the test, then it is returned. Otherwise, a new counterfactual is computed by exploring the neighbourhood of previously generated explanations.

3.3.3 Robustness Against Model Multiplicity

In the previous section we saw how the validity of counterfactual explanations may be affected by changes in the parameters of a machine learning model. In this section we will analyse a closely related scenario, *model multiplicity*, and discuss the challenges that this poses to counterfactual explanations.

Model multiplicity, also known as the Rashomon Effect [35], describes a phenomenon whereby multiple equally accurate models can be trained for a single learning task [36, 37]. Recent results on this problem have shown that model multiplicity may give rise to models that may have distinct behaviours for the same input, both in terms of predicted outcomes and related explanations [38, 39].

Multiplicity can be dealt with by aggregating the outputs that individual models produce for a given input towards a single prediction as suggested in, e.g. [37, 40]. However, the way in which model decisions are aggregated has important implications for counterfactual explanations. Indeed, a counterfactual explanation for an aggregated prediction may not be valid for some (or even all) of the aggregated models, potentially raising concerns about how the counterfactual was chosen. Let us turn to Example 3.4 to understand the ramifications of this problem.

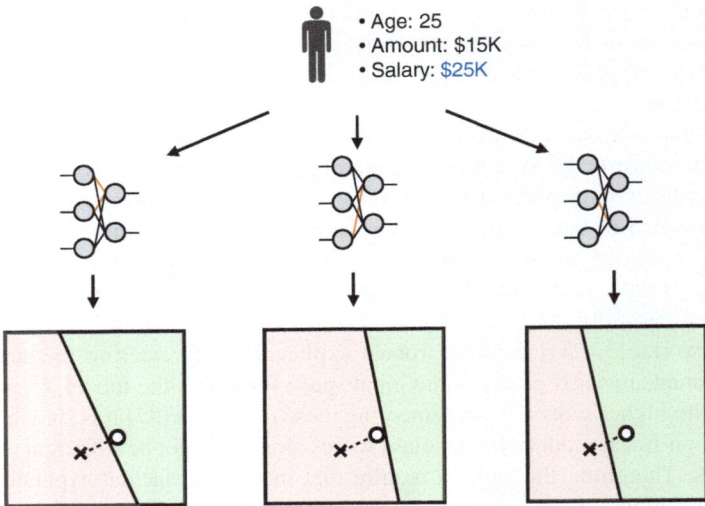

Fig. 3.9 A fictional illustration of model multiplicity, where three neural networks have been trained for the loan application scenario. The three models agree that the original application should be rejected (*times*). However, due to differences in their inner workings, they may not all agree on the magnitude of the salary increase needed for the loan to be granted, thus resulting in the counterfactual explanation (*open circle*) not being valid for all models

Example 3.4 Consider a scenario in which Bob applies for a loan and the bank uses an ensemble of neural networks to predict whether the loan will be repaid, as shown in Fig. 3.9. For the sake of this example, let us assume an ensemble of three neural networks $S = \{f^\theta, f^{\theta'}, f^{\theta''}\}$, where the three models have accuracy of 98.5%, 98.6% and 98.45% respectively. The models exhibit similar performance overall but, due to differences in their internal parameters, produce different predictions for the input $\mathbf{x}$ representing Bob's application, e.g. $f^\theta(\mathbf{x}) = 0$, $f^{\theta'}(\mathbf{x}) = 1$ and $f^{\theta''}(\mathbf{x}) = 0$. Seeing the application rejected, Bob might want to ask for a counterfactual explanation to understand what would need to be changed in the application for it to be accepted next time. However, the disagreement among models in S might pose a challenge to the bank: why would Bob need to change anything when $f^{\theta'}$, which was trained on the same data as f^θ and $f^{\theta''}$ and has comparable accuracy, classified Bob as creditworthy?

Example 3.4 describes only one of many scenarios in which model multiplicity could be problematic. Consider for instance what could happen even if all the three models agreed on the initial prediction. Computing a (nearest) counterfactual explanation for each individual model might result in different recourse recommendations, e.g. suggesting salary increments of $5000, $6000 and $4500 respectively. Given this variability, how is the bank supposed to choose which recommendation should be given to the user? Choosing the least effort recommendation ($4500) might be desirable for Bob, but a higher salary increase might give better assurance

3.3 Robustness of Counterfactual Explanations

to the bank that the loan will be repaid. The variability introduced by model multiplicity is likely to confound the expectations of stakeholders of counterfactual explanations and might raise fairness concerns. Therefore, when generating explanation under model multiplicity one might be interested in taming such variability, which leads to the following high-level definition of robustness:

> Assume an input **x** and a set of equally accurate models S. Let $\mathcal{A}(S, \mathbf{x})$ be an **aggregated prediction** for **x**, and **x**$'$ be a counterfactual explanation for $\mathcal{A}(S, \mathbf{x})$. Robustness against model multiplicity requires that **x**$'$ is a **valid counterfactual explanations for a subset of models** $S' \subseteq S$.

The definition of S' was purposely left unspecified above, as different applications might have different requirements. For instance, when all models in S agree on the initial prediction, one might be interested in computing an explanation that is valid for all models in the set. On the other hand, if disagreement arises within S, one might instead want to find a counterfactual that is valid for the maximum number of non-conflicting models.

Several proposals have been made in the literature to formalise these ideas. For instance, the authors of [41] consider a binary classification setting where all models in S produce the same prediction for an input **x**, i.e. $\mathcal{A}(S, \mathbf{x}) = f^\theta(\mathbf{x})$ for all $f^\theta \in S$ and say that a counterfactual **x**$'$ is robust if it is valid for all models in S. In other words, a counterfactual is robust if it generates the desired outcome across all models in the multiplicity set.

As already noted, satisfying this requirement may be impossible in general due to potential disagreements within S. Thus, alternative characterisations of S' have been proposed in [42]. The authors assume a setting where given an input **x**, a counterfactual explanation is computed for each model in isolation. This results in a set of counterfactuals $C = \{\mathbf{x}'_\theta, \mathbf{x}'_{\theta'}, \mathbf{x}'_{\theta''}, \ldots\}$, where we slightly abuse notation and use $\mathbf{x}'_\theta$ to refer to the counterfactual explanation computed for f^θ. Then, the authors aim at identifying a set of models S' and corresponding explanations from C that satisfy the following properties:

1. *Non-emptiness*, which requires that the set S' must contain at least one model and its corresponding explanation, i.e. $|S'| \neq \emptyset$;
2. *Non-triviality*, extending non-emptiness by asking that S' contains more than just one model, i.e. $|S'| > 1$;
3. *Model agreement*, enforcing that all models contained in S' agree on the initial prediction for **x**, i.e. $\forall f^\theta, f^{\theta'} \in S'. f^\theta(\mathbf{x}) = f^{\theta'}(\mathbf{x})$
4. *Majority vote*, which extends model agreement to ensure that the set S' is the largest possible;
5. *Counterfactual validity*, which enforces that a counterfactual explanation is selected from C only if it is valid for all models in S';
6. *Counterfactual coherence*, which ensures that if a counterfactual explanation is returned from C, then the model for which the explanation was originally generated is also contained in S'.

Besides properties specifically defined for counterfactual explanations, several metrics have been introduced to evaluate model multiplicity in the machine learning literature. Although these metrics were not originally conceived to measure robustness, most of them bear similarities with the properties discussed above. For instance, *arbitrariness* is proposed in [43] to count the number of models within S which give conflicting predictions for the same input **x**. Clearly, the same metric could be lifted to the counterfactual setting, where one could count the number of times for which a counterfactual explanation generated for a single model is valid for other models in S. A variant of the arbitrariness property, called *pairwise disagreement*, has also been studied in [37], where the authors specialise the counting strategy to focus on the number of times in which pairs of models from S disagree.

Having defined model multiplicity and related metrics, we can now discuss existing solutions to generate counterfactual explanations with improved robustness. To the best of our knowledge, the first study of robustness of counterfactual explanations for recourse was presented in [44], where the authors show that explanations that have high plausibility (cf. Sect. 3.2) tend to retain higher validity under model multiplicity. Although the authors did not propose an algorithm to compute more robust explanations, their result suggest that existing procedures to generate counterfactuals that lie within the data manifold could be leveraged to obtain more robust explanations.

The authors of [41] focus on a setting in which S contains feed-forward neural networks that all agree on the initial prediction and propose a construction that combines all networks in S into a single network that is guaranteed to simulate all possible behaviours of the original models. Using this construction, the authors were able to prove that generating a robust counterfactual over a set of neural networks using piece-wise linear activations is NP-complete. Then, extending the standard MILP formulation to counterfactual generation [6], the authors also presented an encoding that is able to generate valid counterfactuals for sets of up to 50 neural networks. The resulting counterfactual explanations showed good plausibility and proximity. However, given the hardness result presented in the paper, the scalability of the approach might be limited by the size of the architectures considered.

In another line of work, techniques from computational argumentation were used in [42] to generate sets of counterfactual explanations satisfying properties (1)–(6) discussed earlier in this section. In a nutshell, the approach consists in clustering similarly-behaving models together to remove conflicts therebetween, to then identify the largest group of models and explanations as the final output.

Other approaches not discussed here, include the approach presented in [45] which extends previous work by the same authors to handle naturally-occurring model changes [24] (cf. Sect. 3.3.2) to the model multiplicity setting. Additionally, ensembling methods such as the one proposed in [40] could be used to address the challenges posed by multiplicity and improve the consistency of predictions produced by disagreeing ensembles.

3.3 Robustness of Counterfactual Explanations

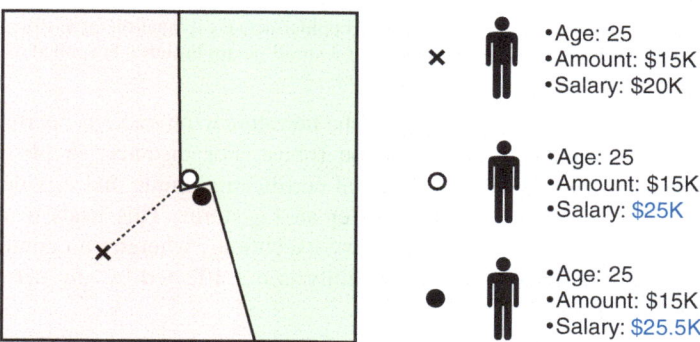

Fig. 3.10 When a counterfactual explanation (*open circle*) lacks robustness to noisy execution, its validity may be compromised if the user does not implement the suggested recourse recommendation to the letter (*filled circle*)

3.3.4 Robustness Against Noisy Execution

The last robustness notion that we will address in this chapter pertains to scenarios in which the validity of a counterfactual explanation may be affected by a noisy implementation of the changes it suggests. Counterfactuals are typically generated under the assumption that the user receiving the explanation will implement the prescribed changes exactly. However, in many cases, it may be impossible to do so, as shown in Example 3.5.

Example 3.5 (Continuing from Example 3.1.) In our running example, Bob receives a counterfactual explanation suggesting that increasing his salary to $25,000 would be sufficient for the loan to be granted. While the recommendation provided by the explanation is clear, achieving the exact increment suggested may not be possible for Bob. For the sake of this example, assume that Bob ends up obtaining a higher salary increment than needed, resulting in an annual salary of $25,500 as shown in Fig. 3.10. Now imagine what would happen if this increment did not result in the loan being granted. Bob may begin to question the extent to which the counterfactual explanation was actually explaining the decision making of the machine learning model, and was not an artefact of the explanation algorithm instead!

Counterfactual explanations are typically obtained by minimising the distance from the input to be explained (cf. Sect. 3.2). As a result, explanations are given by inputs that lie on, or close to, the decision boundary of the machine learning model. This implies that small perturbations to the counterfactual may result in the corresponding input crossing the decision boundary back into the undesired input region, thus compromising the validity of the suggested changes. This phenomenon is aggravated when models have highly non-linear decision boundaries, as it often happens with neural networks. This leads to the following formulation of robustness:

Assume an input **x** and a model f^θ. Let $\mathbf{x}'$ be a counterfactual explanation for **x**. Robustness against noisy execution requires that whenever a **small perturbation** ε is applied to $\mathbf{x}'$, its validity is not affected, i.e. $f^\theta(\mathbf{x}') = f^\theta(\mathbf{x}' + \varepsilon)$.

Different proposals have been made in the literature to characterise perturbations against which a counterfactual should be robust. For instance, in [46, 47] the problem is approached from an adversarial certification angle and ε is defined in terms of bounded perturbations, e.g. under an ℓ_∞ norm. This leads to a worst-case characterisation of robustness to noisy execution, whereby a counterfactual explanation $\mathbf{x}'$ is considered robust if validity is not affected by any perturbation within the ε considered:

$$\forall \mathbf{x}'' \text{ s.t. } \|\mathbf{x}' - \mathbf{x}''\|_p \leq \varepsilon \implies f^\theta(\mathbf{x}') = f^\theta(\mathbf{x}'')$$

for a p-norm $\|\cdot\|_p$ defined as usual. A less conservative proposal is made in [48], where the magnitude of perturbations is allowed to change across different input features based on domain knowledge. A probabilistic approach is taken in [28], where ε is assumed to be drawn from a Gaussian distribution $\mathcal{N}(0, \sigma^2)$. Based on this assumption, the authors develop a metric termed *recourse invalidation rate*, which intuitively captures the expected difference in the predictions of a model f^θ when presented with $\mathbf{x}'$ and its perturbed version $\mathbf{x}' + \varepsilon$ for a random ε, i.e. $\mathbb{E}_\varepsilon[f^\theta(\mathbf{x}') - f^\theta(\mathbf{x}' + \varepsilon)]$.

Existing solutions to generate robust counterfactuals in this setting leverage a variety of techniques, mirroring the different robustness formulations discussed above. For instance, given a counterfactual explanation $\mathbf{x}'$, formal verification procedures are used in [47] to identify regions around $\mathbf{x}'$ for which the classification produced by f^θ does not change when noise is applied. The authors propose a simple binary search procedure, whereby robustness is tested for increasing values of ε; whenever the robustness test fails the procedure terminates and returns the last robust region identified. A robust optimisation approach is instead proposed in [49], where a robust counterfactual $\mathbf{x}'$ computed for an input **x** by solving the following optimisation problem:

$$\underset{\mathbf{x}' \in \mathbb{R}^{n_x}}{\arg\min} \; cost(\mathbf{x}, \mathbf{x}') \text{ s.t. } f^\theta(\mathbf{x}'') = f^\theta(\mathbf{x}'), \forall \mathbf{x}'' \text{ s.t. } \|\mathbf{x}' - \mathbf{x}''\|_p \leq \varepsilon$$

for a given p-norm $\|\cdot\|_p$. A similar approach is taken in [46], where the authors address the robustness problem in a causal setting.

A procedure to generate robust counterfactuals is also given in [28], where Eq. (3.2) is extended with an additional loss term accounting for the invalidations rate of candidate solutions. As a closed-form equation for the invalidation rate may be impossible to obtain for general machine learning models, the authors derive a closed-form expression based on a (linear) local approximation of the model. A similar idea is used in [48], where genetic algorithms are used to solve the resulting optimisation problem.

3.4 Discussion and Future Work

In this chapter we examined different notions of robustness formulated for counterfactual explanations, as well methods for generating robust counterfactual explanations in practice. Through this analysis, we identified open research questions that cut across these robustness notion and methods. This section explores these questions and provides a vision for the future of this rapidly developing field of research.

Research Question 1: One Robustness Notion to Fit Them All? Different approaches to robust counterfactual explanations appear to share similarities, especially when focusing on the same model type. For instance, when dealing with deep neural networks, similar optimisation techniques have been leveraged to handle noise in the explanation generation process. Therefore, interesting connections between different notions of robustness might exist. This has already been observed, e.g. in the context of model changes and model multiplicity [45]; however, we believe further research is needed to explore these relationships.

Research Question 2: Robustness, Fairness and Privacy: Friends or Foes? Besides investigating similarities between different robustness notions, we believe that potentially interesting connections may existing with other important properties studied in machine learning. These include, e.g. fairness and privacy. We have already seen that a lack of robustness to input changes may increase the risk of producing counterfactual explanations that are not fair. Similarly, recent work has exposed privacy risks that are associated with the generation of counterfactuals [50]. Studying whether enforcing robustness may mitigate these risks, or exacerbate them, is therefore a promising research direction.

Research Question 3: Robust Explanations or Robust Models? Another potential area of research lies at the intersection between robustness of explanations and the overall robustness of the underlying model. This is particularly interesting for deep neural networks, given that mainstream approaches to generate counterfactual explanations for these models, e.g. [4], essentially solve an optimisation problem similar to those typically considered in adversarial machine learning [51]. Therefore, one could expect that robust training procedures routinely used in machine learning, e.g. [52], might affect the quality and the robustness of counterfactual explanations.

Research Question 4: Computing Non-adversarial Counterfactuals Counterfactual explanations often lack robustness because the algorithms devised to compute them tend to prioritise the proximity property. However, researchers have shown that this often results in counterfactual explanations that are indistinguishable from adversarial examples [53]. This behaviour is not desirable in many cases, as explanations are meant to capture the decision-making processes of a model, and not its blind spots. Therefore, a promising line of research is to move away from adversarial characterisations of counterfactual explanations and consider instead novel formulations resulting in more informative (and robust) counterfactuals.

Research Question 5: Does Robustness Affect Trust? Robustness is typically conceived as functional property of counterfactual explanations and evaluated using machine-centric metrics. However, a lack of robustness can undermine the credibility and explanatory power of counterfactual explanations. Previous research has not sufficiently investigated this issue through user studies. We argue that such experiments are crucial to understanding the impact of a lack of robustness in counterfactual explanations and should guide the development of new algorithmic approaches.

Research Question 6: A Common Benchmark to Evaluate Robustness When assessing counterfactual explanations, most studies include only a limited set of robust baseline approaches. This is understandable given that most studies on robustness of counterfactuals emerged only within the last couple of years. However, such limited comparisons may hinder our ability to assess and understand the practical utility of different methods. For this reason, we believe that research efforts should be put into developing standardised evaluation frameworks for robustness, perhaps drawing inspiration from similar initiative already undertaken in XAI [54].

References

1. Solon Barocas, Andrew D. Selbst, and Manish Raghavan. The hidden assumptions behind counterfactual explanations and principal reasons. In *FAT* '20: Conference on Fairness, Accountability, and Transparency, Barcelona, Spain, January 27–30, 2020*, pages 80–89, 2020.
2. Ruth M. J. Byrne. Counterfactuals in explainable artificial intelligence (XAI): evidence from human reasoning. In *Proceedings of the Twenty-Eighth International Joint Conference on Artificial Intelligence, IJCAI 2019, Macao, China, August 10–16, 2019*, pages 6276–6282, 2019.
3. Tim Miller. Explanation in artificial intelligence: Insights from the social sciences. *Artif. Intell.*, 267:1–38, 2019.
4. Sandra Wachter, Brent D. Mittelstadt, and Chris Russell. Counterfactual explanations without opening the black box: Automated decisions and the GDPR. *Harv. JL & Tech.*, 31:841, 2017.
5. Vincent Tjeng, Kai Yuanqing Xiao, and Russ Tedrake. Evaluating robustness of neural networks with mixed integer programming. In *7th International Conference on Learning Representations, ICLR 2019, New Orleans, LA, USA, May 6–9, 2019*. OpenReview.net, 2019.
6. Kiarash Mohammadi, Amir-Hossein Karimi, Gilles Barthe, and Isabel Valera. Scaling guarantees for nearest counterfactual explanations. In *AIES '21: AAAI/ACM Conference on AI, Ethics, and Society, Virtual Event, USA, May 19–21, 2021*, pages 177–187. ACM, 2021.
7. Berk Ustun, Alexander Spangher, and Yang Liu. Actionable recourse in linear classification. In *Proceedings of the Conference on Fairness, Accountability, and Transparency, FAT* 2019, Atlanta, GA, USA, January 29–31, 2019*, pages 10–19. ACM, 2019.
8. Amit Dhurandhar, Pin-Yu Chen, Ronny Luss, Chun-Chen Tu, Pai-Shun Ting, Karthikeyan Shanmugam, and Payel Das. Explanations based on the missing: Towards contrastive explanations with pertinent negatives. In *Advances in Neural Information Processing Systems 31: Annual Conference on Neural Information Processing Systems 2018, NeurIPS 2018, December 3–8, 2018, Montréal, Canada*, pages 590–601, 2018.

References

9. Ramaravind Kommiya Mothilal, Amit Sharma, and Chenhao Tan. Explaining machine learning classifiers through diverse counterfactual explanations. In *Proceedings of the ACM Conference on Fairness, Accountability, and Transparency (FAT*'20).*, pages 607–617, 2020.
10. Amir-Hossein Karimi, Gilles Barthe, Bernhard Schölkopf, and Isabel Valera. A survey of algorithmic recourse: Contrastive explanations and consequential recommendations. *ACM Comput. Surv.*, 55(5):95:1–95:29, 2023.
11. Junqi Jiang, Francesco Leofante, Antonio Rago, and Francesca Toni. Robust counterfactual explanations in machine learning: A survey. In *Proceedings of the Thirty-Third International Joint Conference on Artificial Intelligence, IJCAI 2024*, pages 8086–8094. International Joint Conferences on Artificial Intelligence Organization, 8 2024.
12. Leif Hancox-Li. Robustness in machine learning explanations: does it matter? In Mireille Hildebrandt, Carlos Castillo, L. Elisa Celis, Salvatore Ruggieri, Linnet Taylor, and Gabriela Zanfir-Fortuna, editors, *FAT* '20: Conference on Fairness, Accountability, and Transparency, Barcelona, Spain, January 27–30, 2020*, pages 640–647. ACM, 2020.
13. Dylan Slack, Anna Hilgard, Himabindu Lakkaraju, and Sameer Singh. Counterfactual explanations can be manipulated. In *Advances in Neural Information Processing Systems 34: Annual Conference on Neural Information Processing Systems 2021, NeurIPS 2021, December 6–14, 2021, virtual*, pages 62–75, 2021.
14. Hidde Fokkema, Rianne de Heide, and Tim van Erven. Attribution-based explanations that provide recourse cannot be robust. *arXiv preprint arXiv:2205.15834*, 2022.
15. André Artelt, Valerie Vaquet, Riza Velioglu, Fabian Hinder, Johannes Brinkrolf, Malte Schilling, and Barbara Hammer. Evaluating robustness of counterfactual explanations. In *IEEE Symposium Series on Computational Intelligence, SSCI 2021, Orlando, FL, USA, December 5–7, 2021*, pages 1–9. IEEE, 2021.
16. Songming Zhang, Xiaofeng Chen, Shiping Wen, and Zhongshan Li. Density-based reliable and robust explainer for counterfactual explanation. *Expert Syst. Appl.*, 226:120214, 2023.
17. Francesco Leofante and Nico Potyka. Promoting counterfactual robustness through diversity. In *Thirty-Eighth AAAI Conference on Artificial Intelligence, AAAI 2024, Thirty-Sixth Conference on Innovative Applications of Artificial Intelligence, IAAI 2024, Fourteenth Symposium on Educational Advances in Artificial Intelligence, EAAI 2014, February 20–27, 2024, Vancouver, Canada*, pages 21322–21330. AAAI Press, 2024.
18. Riccardo Guidotti, Anna Monreale, Fosca Giannotti, Dino Pedreschi, Salvatore Ruggieri, and Franco Turini. Factual and counterfactual explanations for black box decision making. *IEEE Intell. Syst.*, 34(6):14–23, 2019.
19. Kentaro Kanamori, Takuya Takagi, Ken Kobayashi, and Yuichi Ike. Counterfactual explanation with missing values. *arXiv:2304.14606*, 2023.
20. André Artelt and Andreas Gregoriades. "How to make them stay?": Diverse counterfactual explanations of employee attrition. In *Proceedings of the 25th International Conference on Enterprise Information Systems, ICEIS 2023, Volume 1, Prague, Czech Republic, April 24–26, 2023*, pages 532–538. SCITEPRESS, 2023.
21. Kentaro Kanamori, Takuya Takagi, Ken Kobayashi, and Yuichi Ike. Counterfactual explanation trees: Transparent and consistent actionable recourse with decision trees. In *International Conference on Artificial Intelligence and Statistics, AISTATS 2022, 28–30 March 2022, Virtual Event*, pages 1846–1870. PMLR, 2022.
22. Dan Ley, Saumitra Mishra, and Daniele Magazzeni. GLOBE-CE: A translation based approach for global counterfactual explanations. In *International Conference on Machine Learning, ICML 2023, 23–29 July 2023, Honolulu, Hawaii, USA*, volume 202 of *Proceedings of Machine Learning Research*, pages 19315–19342. PMLR, 2023.
23. Sohini Upadhyay, Shalmali Joshi, and Himabindu Lakkaraju. Towards robust and reliable algorithmic recourse. In *Advances in Neural Information Processing Systems 34: Annual Conference on Neural Information Processing Systems 2021, NeurIPS 2021, December 6–14, 2021, virtual*, pages 16926–16937, 2021.
24. Faisal Hamman, Erfaun Noorani, Saumitra Mishra, Daniele Magazzeni, and Sanghamitra Dutta. Robust counterfactual explanations for neural networks with probabilistic guarantees.

In *International Conference on Machine Learning, ICML 2023, 23–29 July 2023, Honolulu, Hawaii, USA*, pages 12351–12367. PMLR, 2023.
25. Luca Marzari, Francesco Leofante, Ferdinando Cicalese, and Alessandro Farinelli. Rigorous probabilistic guarantees for robust counterfactual explanations. In *Proceedings of the 27th European Conference on Artificial Intelligence, ECAI 2024*, Frontiers in Artificial Intelligence and Applications. IOS Press, 2024.
26. Junqi Jiang, Francesco Leofante, Antonio Rago, and Francesca Toni. Formalising the robustness of counterfactual explanations for neural networks. In *Thirty-Seventh AAAI Conference on Artificial Intelligence, AAAI 2023, Thirty-Fifth Conference on Innovative Applications of Artificial Intelligence, IAAI 2023, Thirteenth Symposium on Educational Advances in Artificial Intelligence, EAAI 2023, Washington, DC, USA, February 7–14, 2023*, pages 14901–14909. AAAI Press, 2023.
27. Emily Black, Zifan Wang, and Matt Fredrikson. Consistent counterfactuals for deep models. In *The Tenth International Conference on Learning Representations, ICLR 2022, Virtual Event, April 25–29, 2022*, 2022.
28. Martin Pawelczyk, Tobias Leemann, Asia Biega, and Gjergji Kasneci. On the trade-off between actionable explanations and the right to be forgotten. In *The Eleventh International Conference on Learning Representations, ICLR 2023, Kigali, Rwanda, May 1–5, 2023*. OpenReview.net, 2023.
29. Sanghamitra Dutta, Jason Long, Saumitra Mishra, Cecilia Tilli, and Daniele Magazzeni. Robust counterfactual explanations for tree-based ensembles. In *International Conference on Machine Learning, ICML 2022, 17–23 July 2022, Baltimore, Maryland, USA*, volume 162 of *Proceedings of Machine Learning Research*, pages 5742–5756. PMLR, 2022.
30. Junqi Jiang, Francesco Leofante, Antonio Rago, and Francesca Toni. Interval abstractions for robust counterfactual explanations. *Artif. Intell.*, 336:104218, 2024.
31. Junqi Jiang, Jianglin Lan, Francesco Leofante, Antonio Rago, and Francesca Toni. Provably robust and plausible counterfactual explanations for neural networks via robust optimisation. In *Asian Conference on Machine Learning, ACML 2023, 11–14 November 2023, Istanbul, Turkey*, volume 222 of *Proceedings of Machine Learning Research*, pages 582–597. PMLR, 2023.
32. Andrea Ferrario and Michele Loi. The robustness of counterfactual explanations over time. *IEEE Access*, 10:82736–82750, 2022.
33. Satyapriya Krishna, Jiaqi Ma, and Himabindu Lakkaraju. Towards bridging the gaps between the right to explanation and the right to be forgotten. In *International Conference on Machine Learning, ICML 2023, 23–29 July 2023, Honolulu, Hawaii, USA*, volume 202, pages 17808–17826, 2023.
34. Ariel Tadeu da Silva and João Roberto Bertini. Using the k-associated optimal graph to provide counterfactual explanations. In *IEEE International Conference on Fuzzy Systems, FUZZ-IEEE 2022, Padua, Italy, July 18–23, 2022*, pages 1–8, 2022.
35. Leo Breiman. Statistical Modeling: The Two Cultures (with comments and a rejoinder by the author). *Statistical Science*, 16(3):199–231, 2001.
36. Charles T. Marx, Flávio P. Calmon, and Berk Ustun. Predictive multiplicity in classification. In *Proceedings of the 37th International Conference on Machine Learning, ICML 2020, 13–18 July 2020, Virtual Event*, volume 119, pages 6765–6774, 2020.
37. Emily Black, Manish Raghavan, and Solon Barocas. Model multiplicity: Opportunities, concerns, and solutions. In *FAccT '22: 2022 ACM Conference on Fairness, Accountability, and Transparency, Seoul, Republic of Korea, June 21–24, 2022*, pages 850–863, 2022.
38. Cynthia Rudin. Stop explaining black box machine learning models for high stakes decisions and use interpretable models instead. *Nat. Mach. Intell.*, 1(5):206–215, 2019.
39. Amanda Coston, Ashesh Rambachan, and Alexandra Chouldechova. Characterizing fairness over the set of good models under selective labels. In *Proceedings of the 38th International Conference on Machine Learning, ICML 2021, 18–24 July 2021, Virtual Event*, volume 139, pages 2144–2155, 2021.

References

40. Emily Black, Klas Leino, and Matt Fredrikson. Selective ensembles for consistent predictions. In *The Tenth International Conference on Learning Representations, ICLR 2022, Virtual Event, April 25–29, 2022*, 2022.
41. Francesco Leofante, Elena Botoeva, and Vineet Rajani. Counterfactual explanations and model multiplicity: a relational verification view. In *Proceedings of the 20th International Conference on Principles of Knowledge Representation and Reasoning, KR 2023, Rhodes, Greece, September 2–8, 2023*, pages 763–768, 2023.
42. Junqi Jiang, Francesco Leofante, Antonio Rago, and Francesca Toni. Recourse under model multiplicity via argumentative ensembling. In *Proceedings of the 23rd International Conference on Autonomous Agents and Multiagent Systems, AAMAS 2024, Auckland, New Zealand, May 6–10, 2024*, pages 954–963. International Foundation for Autonomous Agents and Multiagent Systems / ACM, 2024.
43. Juan Felipe Gómez, Caio Vieira Machado, Lucas Monteiro Paes, and Flávio P. Calmon. Algorithmic arbitrariness in content moderation. In *The 2024 ACM Conference on Fairness, Accountability, and Transparency, FAccT 2024, Rio de Janeiro, Brazil, June 3–6, 2024*, pages 2234–2253. ACM, 2024.
44. Martin Pawelczyk, Klaus Broelemann, and Gjergji Kasneci. On counterfactual explanations under predictive multiplicity. In *Proceedings of the Thirty-Sixth Conference on Uncertainty in Artificial Intelligence, UAI 2020, virtual online, August 3–6, 2020*, volume 124, pages 809–818, 2020.
45. Faisal Hamman, Erfaun Noorani, Saumitra Mishra, Daniele Magazzeni, and Sanghamitra Dutta. Robust algorithmic recourse under model multiplicity with probabilistic guarantees. *IEEE J. Sel. Areas Inf. Theory*, 5:357–368, 2024.
46. Ricardo Dominguez-Olmedo, Amir-Hossein Karimi, and Bernhard Schölkopf. On the adversarial robustness of causal algorithmic recourse. In *International Conference on Machine Learning, ICML 2022, 17–23 July 2022, Baltimore, Maryland, USA*, volume 162, pages 5324–5342, 2022.
47. Francesco Leofante and Alessio Lomuscio. Robust explanations for human-neural multi-agent systems with formal verification. In *Proceeding of the 20th European Conference on Multi-Agent Systems, EUMAS 2023, Naples, Italy, September 14–15*, volume 14282, pages 244–262, 2023.
48. Marco Virgolin and Saverio Fracaros. On the robustness of sparse counterfactual explanations to adverse perturbations. *Artif. Intell.*, 316:103840, 2023.
49. Donato Maragno, Jannis Kurtz, Tabea E. Röber, Rob Goedhart, S. Ilker Birbil, and Dick den Hertog. Finding regions of counterfactual explanations via robust optimization. *arXiv:2301.11113*, 2023.
50. Martin Pawelczyk, Himabindu Lakkaraju, and Seth Neel. On the privacy risks of algorithmic recourse. In *International Conference on Artificial Intelligence and Statistics, 25–27 April 2023, Palau de Congressos, Valencia, Spain*, volume 206 of *Proceedings of Machine Learning Research*, pages 9680–9696. PMLR, 2023.
51. Timo Freiesleben. The intriguing relation between counterfactual explanations and adversarial examples. *Minds Mach.*, 32(1):77–109, 2022.
52. Aleksander Madry, Aleksandar Makelov, Ludwig Schmidt, Dimitris Tsipras, and Adrian Vladu. Towards deep learning models resistant to adversarial attacks. In *6th International Conference on Learning Representations, ICLR 2018, Vancouver, BC, Canada, April 30 - May 3, 2018, Conference Track Proceedings*. OpenReview.net, 2018.
53. Martin Pawelczyk, Chirag Agarwal, Shalmali Joshi, Sohini Upadhyay, and Himabindu Lakkaraju. Exploring counterfactual explanations through the lens of adversarial examples: A theoretical and empirical analysis. In *International Conference on Artificial Intelligence and Statistics, AISTATS 2022, 28–30 March 2022, Virtual Event*, volume 151 of *Proceedings of Machine Learning Research*, pages 4574–4594. PMLR, 2022.

54. Martin Pawelczyk, Sascha Bielawski, Johannes van den Heuvel, Tobias Richter, and Gjergji Kasneci. CARLA: A python library to benchmark algorithmic recourse and counterfactual explanation algorithms. In *Proceedings of the Neural Information Processing Systems Track on Datasets and Benchmarks 1, NeurIPS Datasets and Benchmarks 2021, December 2021, virtual*, 2021.

Chapter 4
Robustness of Saliency-Based Explanations

4.1 Introduction

Salience-based explanations are those which seek to identify the features in an image that are most important to a machine learning model's prediction. In doing so, these explanations reveal important information to human stakeholders (developers, end-users, etc.) about how the machine learning model functions. This information then has the potential to engender trust that models decisions are aligned with how humans would make the given decision [1]. The focus of this chapter is generating salience-based explanations and testing their robustness.

We organise this chapter as follows: we begin by developing a few of the most pertinent salience-based explanation methods on a technical level and establish the different evaluation criteria for these explanation methods. We then lay out the argument for why robustness of explanation is a critical facet without which all utility is lost, and we formulate the robustness of an explanation in a general optimisation framework. We then present the key facets of the literature as methods for solving this optimisation problem demonstrating their strengths and weaknesses illustrated with a set of basic experiments. We conclude by providing a discussion of the broader literature not covered in technical details of this chapter as well as a discussion of the future prospects of the study of robustness in saliency-based explanability.

4.2 Saliency-Based Explanations

Given a machine learning model f^θ which maps inputs $\mathbf{x} \in \mathbb{R}^n$ to outputs $\mathbf{y} \in \mathbb{R}^m$ a saliency based explanation is vector-valued function E that may take as input any number of objects (e.g., input features, model parameters, reference input, etc.) and returns a vector $\mathbf{s} \in \mathbb{R}^n$ where each component $\mathbf{s}_i$ represents the computed salience of the ith feature, x_i. We focus on *local* explanations which are those that are specific to a provided input. The general convention is that positive entries of s indicate features that contributed positively towards or support the given prediction/classification and negative entries indicate that features value does not support the given prediction with higher magnitude representing higher saliency. For unfamiliar readers this convention may reasonably lead to the following questions:

- *Where did this convention and kind of explanation originate?*
- *How can we be sure an explanation,* s *is faithful to the actual model?*

The second question will be the subject of Sect. 4.3 where we will detail computational methods that allow us to approximately understand the fidelity and utility of a given explanation. Conveniently, we are able to provide an answer to the first question while avoiding the mathematical complexities of the second question by introducing one of the most basic "machine learning" models: linear regression. Linear regression is hailed as an interpretable machine learning model, meaning that the relationship between the input features and the output is clear, and the model's parameters can be directly understood as the contribution of each feature to the prediction.

To demonstrate this concept precisely we provide the basic formulation of a linear regression model $f^\theta : \mathbb{R}^n \to \mathbb{R}$ which takes the form:

$$y = \theta^\top x$$

with $\theta \in \mathbb{R}^n$ representing the vector of coefficients, and x represents the vector of input features. Each element θ_i in θ corresponds to a specific feature x_i in x and indicates the strength and direction of the relationship between that feature and the output y. The simplicity of this linear relationship allows for straightforward interpretation: the coefficient θ_i tells us how much the output y is expected to change with a one-unit increase in the corresponding feature x_i, assuming all other features remain constant.

This linear relationship forms the basis for understanding saliency-based explanations. In fact, the coefficients θ_i can be viewed as direct indicators of feature importance or saliency and can be taken directly to be our explanation s for any given input as it satisfies our general notion of salience. Positive values of θ_i imply that an increase in the corresponding feature x_i will result in an increase in the predicted output y, while negative values suggest the opposite effect. The magnitude of θ_i indicates the strength of this relationship: larger magnitudes correspond to more influential features. The fact that linear models allow such straightforward explanations for all of their predictions is why they are considered "interpretable": they are explainable by design. Though the majority of machine learning models do not fit this criterion, linear models serve as a foundational example for more complex models where similar notions of feature importance will be employed.

4.2 Saliency-Based Explanations

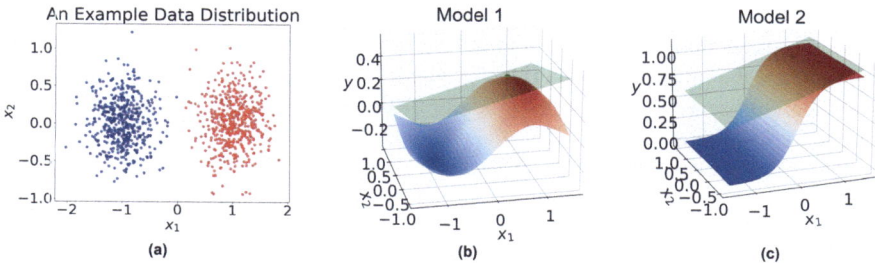

Fig. 4.1 An example case of using gradient-based explanations on a toy dataset. (**a**): We plot the example toy dataset used which contains two groups of points coloured in red and blue that represent two different classes. (**b**): A candidate model using a basis-expanded linear regression model plotted as a surface plot with red and blue coloration indicating the different classes the model assigns. We additionally plot an input point in green and the gradient as a tangent plane representing the input explanation at this point. (**c**): Following the format of subfigure (**b**), we plot a second candidate model using logistic regression with a surface plot along with a green input point and its input gradient as an explanation

4.2.1 The Gradient as an Explanation

Generalising the intuition from linear models to more complex models is done most simply by observing that the input gradient provides the same kind of information we desire for a saliency-based explanation. Recall that the gradient of a function at a point describes the rate of change of the function at that point: it tells us how the function changes with a small change to each feature dimension and thus can be directly interpreted as a saliency-based explanation but for non-linear models. In Fig. 4.1, we provide a visual example of how the gradient can be used as an explanation. In particular in Fig. 4.1b and c, we demonstrate two different non-linear models that predict well on the given task. For both models, we additionally select an input point and plot its location and gradient using a green plane. We can see that this green plane serves as a linear approximation of the function at this point and tells us how changes in the input will approximately affect the model's output.[1] To understand exactly how this works on a technical level, consider the logistic regression model (Model 2 in Fig. 4.1), the variant of linear regression used for classification:

$$y = \sigma(\boldsymbol{\theta}^\top \mathbf{x})$$

where σ is the sigmoid function defined as $\sigma(z) = \frac{1}{1+e^{-z}}$. The sigmoid function maps the linear combination of inputs to a value between 0 and 1, which can be interpreted as a the confidence of the model or in some cases as a probability. For

---

[1] In Chap. 3, where we explore counterfactual explanations, we require that the explanation come with guarantees about how a given change affects the model prediction.

this model, even if it is difficult to see, we can use the same coefficient logic from the previous section. Given that the model's interpretablity is less straight-forward, one may use the more general-purpose gradient-based explanation methods. We can analytically observe that the input gradient of the model provides us with the same salience information.

To make this observation concrete, let's compute the gradient of the output y with respect to the input vector $\mathbf{x}$:

$$\nabla_{\mathbf{x}} y = \frac{\partial y}{\partial \mathbf{x}} = \frac{\partial \sigma(\boldsymbol{\theta}^\top \mathbf{x})}{\partial \mathbf{x}} = \frac{\partial \sigma(z)}{\partial z} \cdot \frac{\partial z}{\partial \mathbf{x}},$$

where $z = \boldsymbol{\theta}^\top \mathbf{x}$. The final gradient can then be obtained by observing the derivative of the sigmoid function with respect to z is $\frac{\partial \sigma(z)}{\partial z} = \sigma(z)(1-\sigma(z))$ and the derivative of z with respect to $\mathbf{x}$ as $\frac{\partial z}{\partial \mathbf{x}} = \boldsymbol{\theta}$. Combining these results, we obtain:

$$\nabla_{\mathbf{x}} y = \sigma(\boldsymbol{\theta}^\top \mathbf{x})(1 - \sigma(\boldsymbol{\theta}^\top \mathbf{x}))\boldsymbol{\theta}.$$

This result shows that the gradient of the output with respect to the input features is proportional to the coefficient vector $\boldsymbol{\theta}$. The sign and relative magnitude of each component of this gradient vector remain intact and therefore reflect the salience or importance of each feature in the input vector $\mathbf{x}$, similar to how the coefficients in linear regression indicate feature importance.

The logistic regression model presented above is notationally very similar to the fully connected neural network architecture we presented in our background section. In fact, logistic regression can be viewed as a zero-hidden-layer neural network. Unfortunately, when we have one or more hidden layers, the same analytical argument above becomes non-trivial as the closed-form of the gradient in terms of the parameters is no longer easy to compute. Yet, for such complicated cases, one can argue that we can still follow the same intuition by observing that the input gradient at a point is a linear approximation of how the function behaves with small changes to the input.

4.2.1.1 Smoothed Gradient

One common issue with directly using the input gradient to explain predictions in deep neural networks is that these gradients can be noisy, particularly when model is highly non-linear. To address this, the concept of a *smoothed gradient* was introduced. The idea behind a smoothed gradient is that for very non-linear functions taking the average of the gradient at several surrounding points may avoid rapidly changing gradients and give us a single, smooth picture of the local function behaviour. Formally, let $\mathbf{x}$ be the original input and let $\mathbf{x}'$ be a perturbed version of $\mathbf{x}$, typically obtained by adding small-magnitude Gaussian noise. The smoothed

4.2 Saliency-Based Explanations

gradient $\tilde{\nabla}_{\mathbf{x}} y$ is then computed as:

$$\tilde{\nabla}_{\mathbf{x}} y = \mathbb{E}_{\mathbf{x}' \sim \mathcal{N}(\mathbf{x}, \sigma^2)} \left[\nabla_{\mathbf{x}'} y \right],$$

where $\mathcal{N}(\mathbf{x}, \sigma^2)$ denotes a Gaussian distribution centred at $\mathbf{x}$ with variance σ^2.

The expectation is typically approximated by averaging the gradients over several perturbed inputs:

$$\tilde{\nabla}_{\mathbf{x}} y \approx \frac{1}{K} \sum_{k=1}^{K} \nabla_{\mathbf{x}'_k} y,$$

where $\mathbf{x}'_k$ are K samples drawn from the distribution $\mathcal{N}(\mathbf{x}, \sigma^2)$. The smoothed gradient often appears to have more regular patterns of salience, particularly in image-based models, where it can highlight regions of an image that are most relevant to a prediction. By averaging out noise, the smoothed gradient approach can be seen as a refinement of the basic gradient method, making it more suitable for explaining complex, non-linear models like deep neural networks.

4.2.1.2 Integrated Gradient

While smoothed gradients offer a way to mitigate noise, they do not address the issue that gradients can be misleading for inputs that lie far from the training distribution or inputs where the model's decision is near the decision boundary. To deal with these concerns, *integrated gradients* were introduced. The idea behind integrated gradients is to compute the cumulative gradient as the input transitions from a baseline (often chosen as a zero vector or a neutral input) to the actual input. This method integrates the gradient of the model's output with respect to the input along a path that linearly interpolates between the baseline and the input.

Formally, let $\mathbf{x}_{\text{baseline}}$ be the baseline input and $\mathbf{x}$ be the actual input. The integrated gradient $\mathbf{g}$ for the i-th feature is given by:

$$\mathbf{g}_i = (\mathbf{x}_i - \mathbf{x}_{\text{baseline},i}) \int_{\alpha=0}^{1} \frac{\partial y(\mathbf{x}_{\text{baseline}} + \alpha(\mathbf{x} - \mathbf{x}_{\text{baseline}}))}{\partial \mathbf{x}_i} d\alpha.$$

This formula essentially sums up the gradients at each point along the straight-line path from the baseline to the actual input, weighted by the distance of the feature from its baseline value. In practice, the salience is taken to be an approximation of this integral given by summing over a discrete set of steps α_k:

$$\mathbf{s}_i = \mathbf{g}_i \approx \sum_{k=1}^{K} \frac{\partial y(\mathbf{x}_{\text{baseline}} + \alpha_k(\mathbf{x} - \mathbf{x}_{\text{baseline}}))}{\partial \mathbf{x}_i} \cdot (\alpha_{k+1} - \alpha_k).$$

Integrated gradients have the appealing property of satisfying several nice logical properties, including sensitivity properties discussed further in [2]. Satisfaction of these logical properties are argued to enable further user trust in the explanations utility.

4.2.2 Perturbation-Based Explanations

In the previous sections, we looked at gradient information as a way to compute explanations from highly non-linear models while maintaining the same general rational as we used for interpretable, linear regression models. One assumption we made throughout our presentation of gradient-based explanations is that we have access to the full model and parameters such that we can compute $\nabla_\mathbf{x} y$, but models are sometimes black-boxes where model owners do not want to share the inner-workings but would like to be transparent in providing an explanation for their prediction. In this case, one may use *perturbation-based* methods to extract an explanation.

To build the perturbation-based explanation framework, we will start by considering black-box computation of a gradient information. The blueprint for this black-box computation is given in the finite difference definition of a gradient: given a function $f : \mathbb{R}^n \to \mathbb{R}$ and a point $\mathbf{x} \in \mathbb{R}^n$, the gradient of f at $\mathbf{x}$ can be approximated using finite differences as follows:

$$\nabla_{\mathbf{x}_i} f(\mathbf{x}) \approx \frac{f(\mathbf{x} + h \cdot \mathbf{e}_i) - f(\mathbf{x})}{h},$$

where $\mathbf{e}_i$ is the i-th standard basis vector (a vector with 1 in the i-th position and 0 elsewhere), and h is a small positive scalar representing the perturbation size. This formula provides an approximation of the partial derivative of f with respect to the i-th feature of $\mathbf{x}$. For the full gradient vector, we apply this formula to each feature individually:

$$\nabla_\mathbf{x} f(\mathbf{x})$$
$$\approx \left[\frac{f(\mathbf{x} + h \cdot \mathbf{e}_1) - f(\mathbf{x})}{h}, \frac{f(\mathbf{x} + h \cdot \mathbf{e}_2) - f(\mathbf{x})}{h}, \ldots, \frac{f(\mathbf{x} + h \cdot \mathbf{e}_n) - f(\mathbf{x})}{h} \right].$$

This finite difference method computes an approximation to the gradient by perturbing each feature of the input $\mathbf{x}$ individually, measuring how the function f changes in response to these small changes. As desired, at no point in the above computations did we assume we had access to the inner-workings of f, only the ability to query its outputs.

However, this method has some limitations. The choice of h is crucial: if h is too large, the approximation may not accurately capture the local behaviour of the function, leading to a poor gradient estimate; if h is too small, numerical precision

4.2 Saliency-Based Explanations

issues may arise, especially in floating-point arithmetic. Moreover, computing the gradient using finite differences can be computationally expensive for high-dimensional inputs, as it requires evaluating the function f at least twice for each dimension.

Yet, there are several potential benefits that are available to use if we abandon computing the gradient as prescribed by the method of finite differences and instead optimise the process for trying to compute an explanation. The seminal work in perturbation-based explanation methods is Local Interpretable Model-Agnostic Explanations (LIME) [3].

4.2.2.1 Local Interpretable Model-Agnostic Explanations (LIME)

Like all of the methods we have seen so far, LIME seeks to generate an explanation for a given input point $\mathbf{x}$. To do so, LIME proposes to sample a series of k perturbation vectors distributed according to $q^{(i)} \sim \mathcal{N}(0, \sigma^2)$. Then it generates predictions for each point:

$$y^{(i)} = f^\theta(\mathbf{x} + q^{(i)})$$

We then take this as a supervised machine learning dataset $\{\mathbf{x} + q^{(i)}, y^{(i)}\}_{i=1}^k$ which we fit a linear model to such that $y^{(i)} \approx \boldsymbol{\beta}^\top(\mathbf{x} + q^{(i)})$. This new linear model given by the vector $\boldsymbol{\beta}$ (and optionally a bias) is then an interpretable model of the function f^θ's local behaviour around $\mathbf{x}$ and thus constitutes an explanation.

Many perturbation-based explanation methods can be formulated as extensions of the LIME framework. One primary modification of LIME that gives rise to new perturbation-based explanations is the use of different local models rather than the linear model proposed above e.g., a quadratic model [4]. In other cases, the perturbation generation is changed in order to adapt the method to domains with different structures i.e., perturbations in the frequency and time space for audio data[5].

4.2.3 Layer-wise Relevance Propagation

While the previous section can be cast as a black-box gradient estimation procedure, another popular approach for explanation generation is to modify the back-propagation process to produce more interpretable explanations. Layer-wise Relevance Propagation (LRP) assumes full access to the model (unlike perturbation-based approaches) but does not compute an input gradient.

Formally, for an input $\mathbf{x} \in \mathbb{R}^n$ and a fully connected neural network with L layers and output $y \in \mathbb{R}$, layer-wise relevance propagation works by redistributing the relevance of the output back through the network to the input features. As in our Chap. 2, let $\mathbf{z}^{(i)}$ denote the activations (post-activations) at layer i, $\mathbf{W}^{(i)}$ the weight matrix connecting layer $i-1$ to layer i, and $\zeta^{(i)} = \mathbf{W}^{(i)}\mathbf{z}^{(i-1)} + \mathbf{b}^{(i)}$ the pre-activation values. The relevance scores, $R_j^{(i)}$, at each neuron j in layer i are propagated backward from the output layer L to the input layer.

At the output layer, the relevance $R^{(L)}$ is initialised as the model output y. This relevance is then propagated back through each layer. For a neuron j in layer i, the relevance is computed as a weighted sum of the relevance of the neurons k in the next layer $(i+1)$, where the weights are determined by the contributions of neuron j to neuron k. This can be expressed mathematically as:

$$R_j^{(i)} = \sum_k \frac{z_j^{(i)} w_{jk}^{(i+1)}}{\sum_{j'} z_{j'}^{(i)} w_{j'k}^{(i+1)}} R_k^{(i+1)},$$

where $z_j^{(i)}$ is the activation of neuron j in layer i, $w_{jk}^{(i+1)}$ is the weight connecting neuron j in layer i to neuron k in layer $(i+1)$, and $R_k^{(i+1)}$ is the relevance of neuron k in layer $(i+1)$. This process continues layer by layer until the relevance scores $R_j^{(0)}$ for the input features are obtained by propagating relevance from the first hidden layer back to the input:

$$\mathbf{s}_j = R_j^{(0)} = \sum_k \frac{x_j w_{jk}^{(1)}}{\sum_{j'} x_{j'} w_{j'k}^{(1)}} R_k^{(1)},$$

where x_j represents the input features. The final relevance scores $R_j^{(0)}$ provide a decomposition of the model's output y into contributions from each input feature.

4.3 Metrics for Saliency-Based Explanations

At a high level, a saliency-based explanation, **s**, provides us with information about the machine learning model. Intuitively, a *good* explanation should accurately reflect how the underlying machine learning model is making its decision. In this section, we focus on quantitative metrics that measure how accurately the saliency map reflects the model's true decision-making process. Ideally an explanations is most accurate if the most salient features identified by the explanation are indeed those that the model relies on most heavily to make a prediction.

Before doing so, we make a critical distinction between the metrics that are presented in this sections and notions of *utility*. In this section, we focus on whether or not an explanation reflects the important features to a model's prediction; however, given that the intended use of explanations for machine learning models is to provide human end-users with valuable information and insights, evaluating the utility or usefulness of an explanation should involve human user-studies.

4.3.1 Fidelity Metrics

One common approach to metric to assess a saliency-based explanation is *fidelity*, sometimes also known as *completeness*. Fidelity is measured by perturbing the input features ranked as most salient and observe the impact on the model's output. Methods involve systematically removing or altering the top-ranked features according to the saliency map and then measuring the resulting change in prediction. If the model's output changes significantly, this suggests that the explanation has correctly identified critical features, thereby demonstrating high fidelity.

Mathematically, consider an input $\mathbf{x}$ and its corresponding saliency map $\mathbf{s}$. Let $f^\theta(\mathbf{x})$ be the original model output, and $f^\theta(\mathbf{x} \setminus \mathbf{x}_{\text{top-k}})$ be the model output after the top-k salient features have been removed or perturbed. A fidelity metric can then be defined as:

$$M_{\text{Fid.}}(\mathbf{s}) = |f(\mathbf{x}) - f(\mathbf{x} \setminus \mathbf{x}_{\text{top-k}})|$$

This metric quantifies the sensitivity of the model's prediction to changes in the features deemed most important by the saliency map. A higher value indicates that the identified features are indeed pivotal to the model's prediction.

4.3.2 Localisation Metrics

Localisation metrics evaluate the spatial or feature-specific accuracy of saliency maps, and require that we have information that certain regions or aspects of the input are known to be relevant to the model's decision. These metrics are especially crucial in fields like computer vision, where explanations are often expected to highlight specific regions within an image that are pertinent to a model's classification decision.

A common localisation metric is Intersection over Union (IoU), which compares the overlap between the saliency map and a ground truth region of interest (ROI). This ground truth might be derived from human annotations or known feature importance. For an image $\mathbf{x}$ with a ground truth region $\mathbf{G}$ and a saliency map $\mathbf{s}$, the IoU is defined as:

$$\text{IoU} = \frac{|\mathbf{G} \cap \mathbf{s}|}{|\mathbf{G} \cup \mathbf{s}|}$$

where $\cap$ denotes the intersection and $\cup$ denotes the union of the ground truth and the salient regions. A higher IoU score indicates that the saliency map accurately localises the features relevant to the model's prediction, closely aligning with human intuition or known regions of importance.

4.4 Robustness of Saliency-Based Explanations

In our previous section, we discussed how to qualitatively assess an explanations *accuracy*, i.e., how well it reflects what is truly important to a model prediction. Further, we made the important distinction between accuracy and *utility* where utility ought to be defined in terms of a human user's ability to glean useful information and therefore enhance their trust in the underlying system. Robustness is a critical property of an explanation without which the accuracy or utility of an explanation are rendered meaningless. In a general sense, the property of robustness for explanations is: small changes to an input or model cause small changes in the explanation.

To demonstrate conceptually why robustness is so crucial, suppose we have a machine learning model f^θ, and input **x**, and an explanation **s**. Further, assume that **s** perfectly represents the importance of each feature (i.e., is highly accurate) and that human users find this information useful (i.e., has high utility). If a small change to the input or model parameter occurs resulting in a new computed explanation **s'** that is entirely different to **s**, then human users who previously found **s** to be useful may now see it as misleading, thus destroying trust in the model and explainability procedure.

We demonstrate this concept pictorially in Figs. 4.2 and 4.3. We began in Fig. 4.1 by demonstrating how for two non-linear models the input gradient serves as a linear approximation of how the function is making its decisions at a point. At a glance, these explanations seemed informative (having high-utility) and faithful (they are truly how the function responds to input changes). In Fig. 4.2, however, we can see that a very small change in the location of our selected input point causes a very

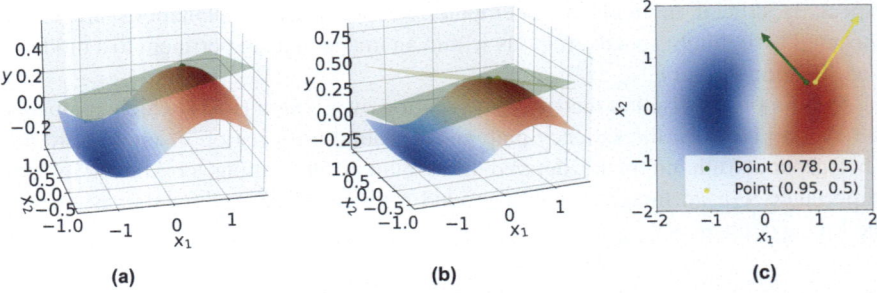

Fig. 4.2 Demonstrating how small input changes can cause large explanation changes. (**a**): The basis-expanded linear regression model plotted as a surface plot with red and blue coloration indicating the different classes the model assigns as in Fig. 4.1. We additionally plot an input point in green and the gradient as a tangent plane representing the input explanation at this point. (**b**): A repeat of subfigure (**a**) but with an additional input point coloured in yellow along with its input gradient which this time is nearly perpendicular to the green input point's gradient. (**c**): We plot the function as a contour as if looking at the functions from above. With each point and its gradient being plotted we can see that a tiny change in the input causes a large change in the input gradient

4.4 Robustness of Saliency-Based Explanations

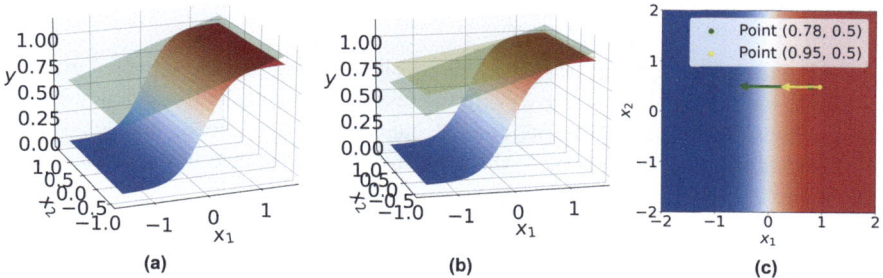

Fig. 4.3 Demonstrating how small input changes can cause large explanation changes. (**a**): The logistic regression model plotted as a surface plot with red and blue coloration indicating the different classes the model assigns as in Fig. 4.1. We additionally plot an input point in green and the gradient as a tangent plane representing the input explanation at this point. (**b**): A repeat of subfigure (**a**) but with an additional input point coloured in yellow along with its input gradient which this time is, unlike in Fig. 4.2, the gradient is largely unchanged. (**c**): We plot the function as a contour as if looking at the functions from above. With each point and its gradient being plotted we can see that for this model, the same kind of input change does not effect such a large explanation change

different gradient to be computed. In fact, the new explanation is nearly orthogonal to the original! Without access to the model, providing these two explanations to end users does the opposite of engendering trust: it allows users to think that the underlying model is chaotic and untrustworthy. Robustness is an attempt to quantify and address this limitation. For example by measuring explanation robustness, we may be guided to select Model 2 over Model 1 as in Fig. 4.3 we can see that the same input change affects only a very small change to the explanation, thus potentially further engendering trust in the underlying system. In this section, we formally define robustness as for saliency-based explanations using the lens of optimisation. We then detail three types of robustness evaluations and describe their benefits and trade-offs.

4.4.1 Defining Robustness for Saliency-Based Explanations

The core principle behind robustness of any system, and explanations are no exception, is that a small change in the system should effect a small change in the output, in this case an explanation. Thus in order to properly defined robustness we have to define what it means to change the system. We focus on two axes: changes to inputs and changes to model parameters. Following this, we provide two perspectives on measuring how much the explanation has changed.

4.4.1.1 Types of Perturbations

Input Perturbations Input perturbations involve making small, controlled modifications to the input data **x** and observing the resulting changes in the saliency-based explanations. These perturbations can be quantified using the ℓ_p metric, where different values of p correspond to different notions of distance in the input space. For example:

- ℓ_0 norm or Hamming distance, captures the number of elements that are different between two vectors
- ℓ_1 norm or Manhattan distance, measures the sum of absolute differences between input features, capturing sparse changes across many features.
- ℓ_2 norm or Euclidean distance, measures the root sum of squared differences, capturing the overall magnitude of changes in the input.
- ℓ_∞ norm considers the maximum change among all input features, focusing on the largest perturbation to any single feature.

For structured inputs, such as text and images, we may consider other domain-specific distance metrics. In Natural Language Processing (NLP), the distance might be measured by the number of words replaced or removed, reflecting syntactic or semantic changes in the text. In the case of images, perturbations might be measured by changes in brightness, hue, or saturation, which correspond to perceptual differences that are more meaningful in the visual domain.

The idea behind these perturbations is to model slight variations that might naturally occur in the input space, such as sensor noise, typographical errors, or slight modifications in real-world data collection. By studying how explanations change under these small perturbations, we can assess the sensitivity of the explanation method to input variations. A robust explanation method should produce similar saliency maps for inputs that are close in the input space, ensuring that the explanations are not overly sensitive to minor, potentially inconsequential changes.

Regardless of how one measures input changes, we will abstractly model the set of inputs that vary according to some *small* changes to an input x with the notation T.

Model Perturbations Model perturbations, on the other hand, involve modifying the parameters of the model, θ, to change the generated explanations. Unlike input perturbations, which are localised and specific to a particular data point, model perturbations have global effects, i.e., potentially altering the explanations for all inputs at once. Of course, in changing the explanations, one also changes the model output. Thus, when crafting a model perturbation an adversary often wants to change the explanation as much as possible while leaving the predictions the same. An example of this kind of optimisation is trying to conceal discriminatory bias in a model where a model might in fact be using protected information (e.g., race, gender, age) to make its predictions, but we would like to modify the model such that explanations show that we are not using these features. We provide an intuitive illustration of this kind of manipulation in Fig. 4.4. The models in a and

4.4 Robustness of Saliency-Based Explanations

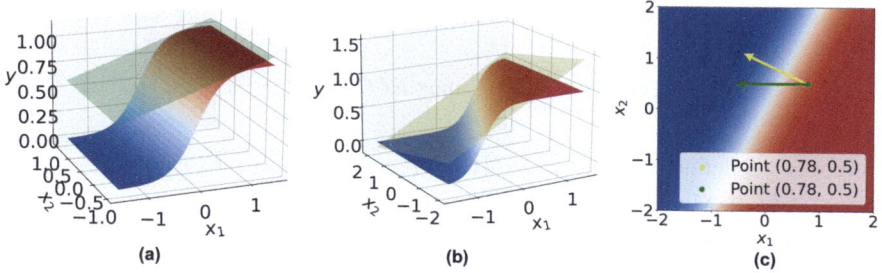

Fig. 4.4 Demonstrating how small model change can cause large explanation change. (**a**): The logistic regression model plotted as a surface plot with red and blue coloration indicating the different classes the model assigns as in Fig. 4.1. We additionally plot an input point in green and the gradient as a tangent plane representing the input explanation at this point. (**b**): Similar to the format of Subfigure (**a**), but this time the model depends slightly on the input feature x_2 rather than solely on x_1 which is represented by a small change to the input weighting of x_2. We can see this alters the function slightly, but also the gradient. (**c**): We plot the altered function as a contour as if looking at the function from above. With each point and its gradient (green being the gradient with respect to the original model) being plotted we can see that with the model change comes an explanation change

b both perform well on the underlying task, visualised in a of Fig. 4.1. However, by changing the function, we are able to considerably alter the explanation at hand. These types of model perturbations are particularly successful in adversarial settings, where an adversary can perturb parameters to find a specific set of model parameters, denoted as θ^{adv}, that systematically corrupts the explanations across different inputs. This has been investigated in [6–8] and in [9, 10] where the authors explore modifying models to conceal discriminatory bias. Similar to the description of T, we use M to denote the set of small changes to model parameters.

4.4.1.2 Bounds on Explanation Robustness

Now that we have formulated what it means to change the system at hand, we now would like to capture the amount of change in the explanation. We start by describing the fragility (the opposite of robustness) of an explanation as the largest difference between the maximum and minimum values of the input gradient for any input inside of the set T and/or any model inside the set M. We then show how one can use the maximum and minimum input gradients to rule out the existence of successful targeted and untargeted manipulations of the explanation. We start by formally stating the key assumptions of our method and then proceed to define definitions for robustness. Now, we proceed define what it means for an explanation to be robust around an input $x \in \mathbb{R}^n$.

Definition 4.1 (Input-Robust Explanation) Given a neural network f^θ, a test input x, a compact input set T, and a vector $\delta \in \mathbb{R}^n$, we say that the explanation of the network f is δ-input-robust iff $\forall i \in [n]$:

$$\left\| \min_{x \in T} E(f, \theta, \mathbf{x})_i - \max_{x \in T} E(f, \theta, \mathbf{x})_i \right\| \leq \delta_i \qquad (4.1)$$

Moreover we define the set $E_T := \{v' \mid \forall i \in [n], \min_{x \in T} E(f, \theta, \mathbf{x})_i \leq v'_i \leq \max_{x \in T} E(f, \theta, \mathbf{x})_i\}$ to be the reachable set of explanations.

An intuitive interpretation of this definition is as a guarantee that for the given neural network, input, and loss, there does not exist an adversary that can change the input gradient to be outside of the set E_T, given that the adversary only perturbs the input by amount specified by T. Further, the vector δ defines the per-feature-dimension fragility (lack of robustness). Under an interval assumption, we take T to be $[x^L, x^U]$ such that $\forall i \in [n], x_i^L \leq x_i \leq x_i^U$. For simplicity, we describe this interval with a width vector $\epsilon \in \mathbb{R}^n$ such that $[x^L, x^U] = [x - \epsilon, x + \epsilon]$. We now present the analogous definition for the model perturbation case.

Definition 4.2 (Model-Robust Explanation) Given a neural network f^θ, a test input x, a compact parameter set M, and a vector $\delta \in \mathbb{R}^n$ we say that the explanation of the network f is δ-model-robust iff $\forall i \in [n]$:

$$\left\| \min_{\theta \in M} E(f, \theta, \mathbf{x})_i - \max_{\theta \in M} E(f, \theta, \mathbf{x})_i \right\| \leq \delta_i. \qquad (4.2)$$

Moreover, we define $E_M := \{v' \mid \forall i \in [n], \min_{\theta \in M} E(f, \theta, \mathbf{x})_i \leq v'_i \leq \max_{\theta \in M} E(f, \theta, \mathbf{x})_i\}$.

This definition is identical to Definition 4.1 save for the fact that here the guarantee is over the model parameters. As such, a model that satisfies this definition guarantees that for the given neural network and input, there does not exist an adversary that can change the explanation outside of E_M given that it only perturbs the parameters by amount specified by M. Similar to what is expressed for inputs, we express M as an interval over parameter space parameterised by a width vector $\gamma \in \mathbb{R}^{n_p}$. To avoid issues of scale, we express an interval over a weight vector W w.r.t. γ as $[W - \gamma \cdot |W|, W + \gamma \cdot |W|]$.

4.4.1.3 Worst-Case Explanations

The above definitions of explanation robustness (Definitions 4.1 and 4.2) require the computation of bounds that rule out two explanations being vastly different for similar inputs. However, bounds may not necessarily be the most informative or desirable way to capture the change in explanation. A different perspective that may be more a intuitive way to rule out the existence of vastly different explanations for similar inputs is searching for *the most different* explanation that results from a similar input. Typically, this is done by defining a function $h : \mathbb{R}^n \times \mathbb{R}^n \to \mathbb{R}$ where

4.4 Robustness of Saliency-Based Explanations

$h(v, v')$ is a function that measures the similarity of v and v', satisfying the standard requirements of a distance function:

- **Non-negativity:** $h(v, v') \geq 0$ for all $v, v' \in \mathbb{R}^n$, and $h(v, v') = 0$ iff $v = v'$.
- **Symmetry:** $h(v, v') = h(v', v)$ for all $v, v' \in \mathbb{R}^n$.
- **Triangle inequality:** $h(v, v'') \leq h(v, v') + h(v', v'')$ for all $v, v', v'' \in \mathbb{R}^n$.

Provided we have access to such a distance function (typical choices include the mean squared error, cosine similarity, and mean absolute error), there are two primary ways optimisation formulations that are used: *targeted* an *untargeted* explanation robustness. There are ways of moving between this definition and the bounds definition above that we discuss later in the chapter. Here, we formally define these two below:

Definition 4.3 (Untargeted Explanation Robustness) Given a neural network f^θ, similarity function h, threshold τ, a test input x, and a compact input set T we say that the explanation of the network f is τ-input-robust iff $\forall x' \in T$:

$$h\big(E(f, \theta, \mathbf{x}), E(f, \theta, \mathbf{x}')\big) \leq \tau \tag{4.3}$$

The intuitive interpretation of Definition 4.3 is that we consider the explanation for a given input to be robust if there are no inputs inside a predetermined set of similar inputs T that cause the explanation to change by more than τ. In later sections, we will see how we can reason computationally about this definition. Before doing so, we provide the targeted definition of explanation robustness:

Definition 4.4 (Targeted Explanation Robustness) Given a neural network f^θ, similarity function h, target explanation v_{targ}, threshold τ, a test input x, and a compact input set T we say that the explanation of the network f is τ-input-robust iff $\forall x' \in T$:

$$h\big(E(f, \theta, \mathbf{x}'), v_{\text{targ}}\big) \geq \tau \tag{4.4}$$

The targeted explanation robustness defined in Definition 4.4 offers a complementary perspective to the untargeted robustness concept. Here, the focus shifts from ensuring that all explanations within a set of similar inputs remain close to the original explanation to ensuring that all such explanations remain far from a specific target explanation v_{targ}. This concept is particularly relevant when evaluating the robustness of explanations against adversarial perturbations, where an adversary might attempt to manipulate the input in such a way that the resulting explanation closely resembles a predetermined, potentially misleading, target explanation.

Together, the untargeted and targeted robustness definitions provide a comprehensive framework for assessing the reliability of explanation methods. While untargeted robustness focuses on general stability across similar inputs, targeted robustness ensures that explanations remain distinct from specific undesirable outcomes. In subsequent sections, we will delve into the computational strategies for estimating and verifying these robustness properties, enabling a more nuanced understanding of the strengths and limitations of various explanation techniques.

4.5 Estimating Robustness with Random Noise

The simplest approach to estimating the robustness of a given example is to sample random perturbations and observe how much the explanation changes. To do so, one simply defines a random variable with support on the specified sets T and M. We can randomly sample points from the defined random variable and evaluate the input gradients at these points. By aggregating these sampled gradients, we can estimate the extremal values required for computing robustness. Below, we state how this looks mathematically for input and model perturbations and then we discuss the strengths and weaknesses of this approach below.

4.5.1 Estimating Input Robustness via Sampling

Estimating Bounds on Input Robustness For an input set T defined as the interval $T = [\mathbf{x} - \epsilon, \mathbf{x} + \epsilon]$, we aim to estimate the maximum and minimum values of the explanation $E(f, \theta, \mathbf{x})$ for any $\mathbf{x} \in T$. Specifically, from Definition 4.1, we want to approximate:

$$\min_{\mathbf{x} \in T} E(f, \theta, \mathbf{x})_i \quad \text{and} \quad \max_{\mathbf{x} \in T} E(f, \theta, \mathbf{x})_i$$

for each feature dimension i. To do this, we can sample N points $\mathbf{x}_1, \mathbf{x}_2, \ldots, \mathbf{x}_N$ uniformly from within the interval T, and compute the explanation $E(f, \theta, \mathbf{x}_j)$ at each sampled point $\mathbf{x}_j$. The sampled estimations of the minimum and maximum explanations are then:

$$\min_{1 \leq j \leq N} E(f, \theta, \mathbf{x}_j)_i \quad \text{and} \quad \max_{1 \leq j \leq N} E(f, \theta, \mathbf{x}_j)_i$$

These sampled minima and maxima provide an approximation of the extremal values over the entire set T. As N increases, these approximations become more accurate, and the difference between the maximum and minimum values provides an estimate of the fragility of the explanation.

Estimating Worst-Case Input Robustness The above estimation of upper and lower bounds, while in line with Definition 4.1, is not in line with how the literature suggests estimating explanation robustness with sampling. Rather, a key assumption the literature makes is the existence of a distance metric $d : \mathbb{R}^{n \times n} \to \mathbb{R}$ which determines the distance between two explanations. Then, by sampling within T, we would like to know which single input causes the largest distance in explanations. To do so, we again assume a random variable with support on the interval $T = [\mathbf{x} - \epsilon, \mathbf{x} + \epsilon]$ and seek to estimate the worst-case fragility by identifying the maximum distance between the explanation at the original input $\mathbf{x}$ and any sampled input within T. Let $E(f, \theta, \mathbf{x})$ be the explanation at the original input $\mathbf{x}$. We randomly sample N points $\mathbf{x}_1, \mathbf{x}_2, \ldots, \mathbf{x}_N$ uniformly from within T and compute the explanations $E(f, \theta, \mathbf{x}_j)$ for each sampled point $\mathbf{x}_j$. The worst-case fragility is then estimated by finding the maximum distance between the original explanation and the sampled explanations:

$$\max_{1 \leq j \leq N} d\left(E(f, \theta, \mathbf{x}), E(f, \theta, \mathbf{x}_j)\right)$$

This estimate gives us the worst-case change in the explanation due to input perturbations within the interval T. As N increases, the estimate becomes more reliable, providing a practical way to assess the fragility of explanations to input perturbations.

4.5.2 *Estimating Model Robustness via Sampling*

Estimating Bounds on Model-Robustness Similarly, for the model robustness, where the parameter set M is defined as $M = [\theta - \gamma \cdot |\theta|, \theta + \gamma \cdot |\theta|]$, from Definition 4.2, we aim to estimate:

$$\min_{\theta \in M} E(f, \theta, \mathbf{x})_i \quad \text{and} \quad \max_{\theta \in M} E(f, \theta, \mathbf{x})_i$$

For this, we sample N parameter vectors $\theta_1, \theta_2, \ldots, \theta_N$ uniformly from within the interval M, and compute the explanation $E(f, \theta_j, \mathbf{x})$ for each sampled parameter vector θ_j. The sampled estimations are:

$$\min_{1 \leq j \leq N} E(f, \theta_j, \mathbf{x})_i \quad \text{and} \quad \max_{1 \leq j \leq N} E(f, \theta_j, \mathbf{x})_i$$

Again, these sampled extrema provide an approximation of the true minima and maxima, and their difference provides an estimate of the fragility of the explanation with respect to model perturbations.

Estimating Worst-Case Model-Robustness As with input robustness, we may be interested in identifying the single worst-case parameter rather than bounds on the parameter. We follow an identical procedure to input robustness defining a random variable with support over $M = [\theta - \gamma \cdot |\theta|, \theta + \gamma \cdot |\theta|]$, and we want to estimate the worst-case change in the explanation according to a distance function h. Let $E(f, \theta, \mathbf{x})$ be the explanation at the original parameters θ. We randomly sample N parameter vectors $\theta_1, \theta_2, \ldots, \theta_N$ uniformly from within M and compute the explanations $E(f, \theta_j, \mathbf{x})$ for each sampled parameter vector θ_j. The worst-case fragility is then estimated by finding the maximum distance between the original explanation and the sampled explanations:

$$\max_{1 \leq j \leq N} h\left(E(f, \theta, \mathbf{x}), E(f, \theta_j, \mathbf{x})\right)$$

This estimate provides the worst-case change in the explanation due to perturbations in the model parameters within the interval M.

4.5.3 Discussion of the Sampling Approach

In this section, we list the advantages and disadvantages of the sampling-based approach to reasoning about explanation robustness. A key advantage is the model-agnostic nature of the approach. No matter what machine learning model, f^θ, is used, the approach of sampling and building bounds still applies. Additionally, the sampling based approach can be done in a black-box setting i.e., without access to the model or its inner workings. A mild drawback of this approach is its computational complexity. For exceedingly large models such as modern foundation models, performing a single inference is expensive. Evaluating robustness with a statistical approach will require potentially thousands of samples which may be time-prohibitive if model inference is not easily parallelizable. Finally, the major drawback of all sampling-based approaches to solving optimisation problems is that as the input dimension grows large, it is highly unlikely that the maximum or minimum value points are sampled. The implication of this is that as the dimensionality of the input grows, so does the under-estimation of the worst-case.

4.6 Auditing Saliency-Based Explanations with Counter Examples

As the input dimension of a machine learning grows, the statistical approach becomes less effective due to the fact that the chance of sampling the input point or model parameter that solves the optimisation problems stated in Definitions 4.1–4.4 vanishes. Unsurprisingly, tools and methods from the field of optimisation do not necessarily suffer from this same issue. Indeed, one of the most seminal papers in robustness of saliency-based explanations, [11], uses a very simple optimisation approach. In this section, we will discuss this work in detail along with its generalisation to weight-space perturbations [6]. Both of these methods deal exclusively with computing worst-case definitions of explanation robustness (i.e., Definitions 4.3 and 4.3).

4.6 Auditing Saliency-Based Explanations with Counter Examples

4.6.1 Input Attacks on Explanation Robustness

Searching for a worst-case input to a model is generally called an *attack*. An attack consists of crafting an input to the system that yields undesirable behaviour. In the case of explanations the undesirable behaviour is dictated by the function h and the threshold τ as laid out in Definitions 4.3 and 4.4. In particular, for untargeted attacks on an explanation given for an input x, we have the following constrained optimisation problem:

$$x_{\text{adv}} \approx \arg\max_{x' \in T} h\left(E(f, \theta, \mathbf{x}), E(f, \theta, \mathbf{x}')\right) \quad (4.5)$$

The goal of the attack is to find a value approximately solving the above optimisation problem such that:

$$h\left(E(f, \theta, \mathbf{x}), E(f, \theta, \mathbf{x}_{\text{adv}})\right) > \tau.$$

If one has indeed found a point $\mathbf{x}_{\text{adv}} \in T$ such that the above inequality holds, then $\mathbf{x}_{\text{adv}}$ serves as proof by counter example that Definition 4.3 does not hold for the given input, model, and explanation method. Thus, it is safe to say that this explanation is not robust. The opposite, unfortunately, does not hold. That is, if we are unable to find an input $\mathbf{x}_{\text{adv}}$ that serves as a counter example to Definition 4.3, we cannot soundly reason that no such example exists. This will be discussed in detail in Sect. 4.7.

For targeted robustness, a very similar process is followed with the optimisation problem being modified to accommodate the new goal of finding an explanation as close to the target explanation v_{targ} as possible:

$$x_{\text{adv}} \approx \arg\min_{x' \in T} h\left(v_{\text{targ}}, E(f, \theta, \mathbf{x}')\right) \quad (4.6)$$

Here, as above, we would like to check that the input found $\mathbf{x}_{\text{adv}}$ satisfies the opposite of the inequality given in Definition 4.4:

$$h\left(v_{\text{targ}}, E(f, \theta, \mathbf{x}')\right) < \tau.$$

This then implies that Definition 4.4 does not hold and that for the given input, model, and explanation method the explanation is not robust.

Variants on the Input-Attack Problem While the formulations we provide above make up the core principles of input attacks on saliency-based explanations, there are several variants that have been discussed in the literature. In both [12] and [11] the authors study the problem of finding a worst-case explanation using perturbations and add the further constraint that the model output should be the same between the perturbed input and original input. In [13] they do not include this constraint, but are not particularly focused on attacks.

Solving the Optimisation Problem The most common method for crafting attacks, employed by [12] and [11] and following the example set in the study of adversarial robustness [14], is the use of first-order optimisation. This consists of using the set of allowable perturbations (defined in this work as T) and using the gradient of an objective function, usually just maximising the value of h, to search T.

4.6.2 Model Attacks on Explanation Robustness

As with the definitions themselves, attacks in the space of model parameters largely follow the same formulation as input attacks. For completeness, we use this section to state the optimisation problems and counter-example criteria. The main focus of this section, however, is a discussion of the relevant literature and coverage of technical details of practical attacks on the model robustness of explanations.

For untargeted attacks on an explanation given an input x, we have the following constrained optimisation problem:

$$\theta_{\text{adv}} \approx \arg\max_{\theta' \in M} h\big(E(f, \theta, \mathbf{x}), E(f, \theta', \mathbf{x})\big) \tag{4.7}$$

The goal of the attack, as with the input attack, is to find a value approximately solving the above optimisation problem such that:

$$h\big(E(f, \theta, \mathbf{x}), E(f, \theta_{\text{adv}}, \mathbf{x})\big) > \tau.$$

Where if a θ_{adv} exists satisfying the above inequality, we know that the model does not satisfy explanation robustness for the given point $\mathbf{x}$.

The corresponding definition holds for targeted attacks. Namely:

$$\theta_{\text{adv}} \approx \arg\min_{\theta' \in M} h\big(v_{\text{targ}}, E(f, \theta', \mathbf{x})\big) \tag{4.8}$$

The property of model-robustness of the explanation at the point x is then violated if one can find a value θ_{adv} such that:

$$h\big(v_{\text{targ}}, E(f, \theta_{\text{adv}}, \mathbf{x})\big) < \tau.$$

Variants on the Model-Parameter Attack Problem Similar to input attacks, several approaches exist for attacking explanations by perturbing model parameters, but these often involve creative manipulation of the model's behaviour rather than simple bounded changes. In [6], the authors explore the problem of finding a set of model parameters that result in misleading explanations while keeping the model's predictions unchanged. This is framed as an optimisation problem aimed at maximising discrepancies in explanations caused by subtle modifications to the model weights and is largely accomplished with gradient-based methods similar to the approaches discussed in our input manipulation section. Other works explore different kinds of manipulations. For instance, [7] introduces a "scaffolding" technique, where adversarial data points are added to the training set. This causes the model to behave normally on the standard data-manifold while behaving adversarially off-manifold which is precisely where explanation methods like LIME and SHAP draw their information. This attack does not correspond to a simple bounded perturbation of the original model parameters but rather introduces entirely new dynamics to the model's behaviour. Similarly, in the domain of "fairwashing," [15] discusses manipulating the feature space by introducing superfluous features. These features allow the classifier to find a set of weights that according to the explanations ignore sensitive attributes, while using them in reality. This shows how non-robust explanations can be leveraged by adversaries to conceal discriminatory biases. As before, this sort of manipulation is not easily characterised by a bounded perturbation to the model's original parameters; instead, it fundamentally alters the feature space in a way that misleads the explanations, making the model appear fair when it is not.

4.7 Proving Robustness of Saliency-Based Explanations

Given a model, input, and explanation method, attacks search for counter-examples that are able to falsify various properties of explanation robustness (i.e., Definitions 4.1–4.4). Yet, when we search with a gradient-based method and do not find a valid counter-example to a property, we cannot say for certain that the property holds. In this section, we focus our attention on computational methods that *are* able to formally prove that a given robustness property of a saliency-based explanation holds. We will begin with a conceptual discussion emphasising what we aim to *prove* about the robustness of saliency-based explanations and will lay down the technical foundations for how this is proven. We will then provide the detailed computations involved in proving the robustness of an explanation for neural network models. As with our previous sections, we end with a discussion of the benefits and drawbacks of this approach to robustness of explainability.

Before proceeding, we highlight that the discussion and methods in this section focus exclusively on using the raw gradient as an explanation. In many cases, it is straight-forward to adapt this certification (i.e., smoothed gradients, integrated gradients); in other cases, it is less obvious. In particular, the below certification does prove the robustness of saliency based explanations based on LIME provided that the sampling scheme for LIME has support that is roughly equal to T, we will clarify this technically in our future discussion. We also remark here on the choice of $\mathcal{L}$ which is used to compute the gradient. Typically, one takes the input-gradient of the predicted class to be the explanation; however, one may want to generate the class-wise explanation for not just the predicted class, but other classes as well [16, 17]. In addition, adversarial robustness literature often considers attacks with respect to surrogate loss functions as valid attack scenarios [18]. By keeping $\mathcal{L}$ flexible in our formulation, we model each of these possibilities.

4.7.1 Certification via Valid Gradient Bounds

We begin by assuming we have an input $\mathbf{x}$ and a set built around this input with a vector of non-negative values denoted $\varepsilon \in \mathbb{R}^{n,+}$ such that we have $T := [\mathbf{x} - \varepsilon, \mathbf{x} + \varepsilon]$. Though we use this construction for our input set, called an interval, more expressive input sets can be used i.e., zonotopes and polytopes [19]. In this simplified exposition, we will take the explanation at this point to be simply the gradient i.e., $E(f, \theta, \mathbf{x}) = \nabla_{\mathbf{x}} f^\theta(\mathbf{x})$. Provided this setting, input robustness (laid out in Definitions 4.1 and 4.3, Definition 4.4), proving explanation robustness requires us to prove that:

$$\forall \mathbf{x}' \in T, \left\| \nabla_{\mathbf{x}'} f^\theta(\mathbf{x}') - \nabla_{\mathbf{x}} f^\theta(\mathbf{x}) \right\| \leq \delta \qquad (4.9)$$

for a given δ where the inequality must hold element-wise. We have seen that if the gradient descent algorithm finds an input that violates the above inequality, then we know it does not hold. However, if we are unable to find an input that is a counter-example we are unable to conclusively reason that the explanation is robust at this point. While we discuss this fact in the context of optimisation in the above section, here we opt to discuss this in the context of logic.

Complete Algorithms Logically speaking, the counter-example/attack algorithms discussed in Sect. 4.6 comprise a *complete* approach to reasoning about the predicate in eq. (4.9). A complete algorithm is one that never answers with false negatives but is allowed to answer with false positives. Thus, a trivially sound algorithm is to always return a positive result. It is clear that if an attack produces a negative answer, then it has found a counter example which can be easily checked. So no negative answer will ever be a false negative. On the other hand, as we discussed in the context of optimisation in the previous section, if we are unable to find a counter example it does not mean that one does not exist. Thus, the algorithm may release false positives: declaring an explanation robust if it is not.

Sound Algorithms Unlike complete algorithms, a sound algorithm never releases false positives but may release false negatives. In the context of measuring the robustness of an explanation, this means that if a sound algorithm returns that a given explanation is robust then we can be certain that no counter example exists. Thus, a positive result returned from a sound algorithm constitutes a *guarantee* that the given explanation is robust. For safety-critical applications, such those in financial services and medical imaging, having such a guarantee provides a solid foundation for trust.

4.7 Proving Robustness of Saliency-Based Explanations

With the goal of this section being to develop a sound approach to guaranteeing explanation robustness, we now lay down the foundation of such a guarantee which is the computation of valid gradient bounds. Formally:

Definition 4.5 (Valid Gradient Bounds) Given a neural network f^θ and an input set T, a pair of vectors $v^L \in \mathbb{R}^n$, $v^U \in \mathbb{R}^n$ is a valid gradient bound if and only if:

$$\forall \mathbf{x}' \in T, v^L \leq \nabla_{\mathbf{x}'} f^\theta(\mathbf{x}') \leq v^U$$

Similarly, we say that the vectors are valid gradient bounds for a parameter set M if and only if:

$$\forall \theta' \in M, v^L \leq \nabla_{\mathbf{x}} f^{\theta'}(\mathbf{x}) \leq v^U$$

Provided that we have access to vectors satisfying the definition of valid gradient bounds we can rule out particular targeted and untargeted attacks:

Lemma 4.1 (Provable Untargeted Attack Robustness) *Given valid parameter bounds* $\mathbf{v}^L \in \mathbb{R}^n$, $\mathbf{v}^U \in \mathbb{R}^n$, *similarity metric h, and success threshold* τ, *one can prove the non-existence of successful untargeted attacker if:*

$$\forall \mathbf{v} \in [\mathbf{v}^L, \mathbf{v}^U], h\left(\nabla_{\mathbf{x}} f^\theta(\mathbf{x}), \mathbf{v}\right) \leq \tau$$

Similarly, for targeted attacks:

Lemma 4.2 (Provable Targeted Attack Robustness) *Given valid parameter bounds* $\mathbf{v}^L \in \mathbb{R}^n$, $\mathbf{v}^U \in \mathbb{R}^n$, *similarity metric h, target explanation* $\mathbf{v}_{targ}$, *and success threshold* τ, *one can prove the non-existence of successful targeted attacker if:*

$$\forall \mathbf{v} \in [\mathbf{v}^L, \mathbf{v}^U], h\left(\mathbf{v}_{targ}, \mathbf{v}\right) \geq \tau$$

The above Lemmas demonstrate how valid gradient bounds can imply the robustness of a gradient-based explanation according to the various definitions of robustness. However, for saliency-based explanations that are not gradient-based they valid gradient bounds may still provide a guarantee of robustness. To demonstrate this, we consider LIME where the perturbed samples on which the local model is trained are drawn from a distribution with support only on a subset of T and where we use a linear model. In this case, we know the fit of the local explainable model to the true gradient at the point $\mathbf{x}$ has error bounded by a value proportional to δ. If δ is 0, for example, then we know the local linear model for $\mathbf{x}$ will have a parameter that is exactly equal to $\nabla_{\mathbf{x}} f^\theta(\mathbf{x})$ and is robust to any change in $\mathbf{x}$ that keeps the perturbation distribution support in T.

4.7.2 Computation of Valid Gradient Bounds

Now that we have shown how valid gradient bounds allow us to compute sound guarantees for the robustness of a given explanation, in this section we provide the technical details on how to compute valid gradient bounds. We first discuss how to compute forward pass bounds with respect to either input intervals or weight intervals (or both) forward through the neural network. We then discuss how to pass the computed bounds backwards through the automatic differentiation process in order to get sound upper and lower bounds on the input gradient that constitute valid gradient bounds.

Forward Pass Bounds Solving problems of the form: $\min \{\cdot \mid x^\star \in [x - \varepsilon, x + \varepsilon]\}$ has been well-studied in the context of adversarial robustness certification. However, optimising over both inputs and parameters, i.e.,

$$\min \left\{ \cdot \mid x^\star \in [x - \varepsilon, x + \varepsilon], \theta^\star \in [\theta^L, \theta^U] \right\}$$

is much less well-studied, and to-date has appeared primarily in the certification of probabilistic neural networks. Existing solutions typically leverage double interval matrix multiplication to bound the output of each layer of the neural network.

Theorem 4.1 (Double Interval Matrix Multiplication) *Given element-wise intervals over matrices* $[\mathbf{A}_L, \mathbf{A}_U]$ *where* $\mathbf{A}_L, \mathbf{A}_U \in \mathbb{R}^{n \times m}$ *and* $[\mathbf{B}_L, \mathbf{B}_U]$ *where* $\mathbf{B}_L, \mathbf{B}_U \in \mathbb{R}^{m \times k}$, *define the matrices* $\mathbf{A}_\mu = (\mathbf{A}_U + \mathbf{A}_L)/2$ *and* $\mathbf{A}_r = (\mathbf{A}_U - \mathbf{A}_L)/2$. *Allow* $\mathbf{B}_\mu$ *and* $\mathbf{B}_r$ *to be defined analogously, then computing using Rump's algorithm [20],*

$$\mathbf{C}_L = \mathbf{A}_\mu \mathbf{B}_\mu - |\mathbf{A}_\mu| \mathbf{B}_r - \mathbf{A}_r |\mathbf{B}_\mu| - \mathbf{A}_r \mathbf{B}_r$$
$$\mathbf{C}_U = \mathbf{A}_\mu \mathbf{B}_\mu + |\mathbf{A}_\mu| \mathbf{B}_r + \mathbf{A}_r |\mathbf{B}_\mu| + \mathbf{A}_r \mathbf{B}_r,$$

we have that $\mathbf{C}_{Li,j} \leq [\mathbf{A}'\mathbf{B}']_{i,j} \leq \mathbf{C}_{Ui,j} \; \forall \mathbf{A}' \in [\mathbf{A}_L, \mathbf{A}_U], \; \mathbf{B}' \in [\mathbf{B}_L, \mathbf{B}_U]$. *Diep [21] showed that the above bounds have a worst-case overestimation factor of 1.5.*

Backwards Pass Bounds In order to compute the input gradient of our model's loss or output, we cannot simply stop at the loss or output, instead we must continue to propagate the interval through the back-propagation process (the usual gradient computations). This involves solving an optimisation problem of the form:

$$\delta_L, \delta_U = \min \& \max \left\{ \partial l / \partial \theta^\star \mid \partial l / \partial y^\star \in [\partial l_L, \partial l_U], \theta^\star \in \left[\theta^L, \theta^U\right], \hat{z}^{(k)} \in \left[\hat{z}_L^{(k)}, \hat{z}_U^{(k)}\right] \right\}$$

The bounds on the intermediate activations ($\hat{z}_L^{(k)}$ and $\hat{z}_U^{(k)}$) are obtained using the interval bound propagation defined in the prior subsection, and the bounds on the partial derivative $[\partial l_L, \partial l_U]$ are derived from the loss function (see discussion in ?).

4.8 Literature Review

We use the double matrix interval arithmetic defined in Theorem 4.1 to propagate intervals (bolded as before) through the backwards pass. Specifically, we can backpropagate $\partial \mathcal{L}/\partial \hat{z}^{(K)} = [\partial l_L, \partial l_U]$ to obtain

$$\frac{\partial \mathcal{L}}{\partial z^{(k-1)}} = \left(W^{(k)}\right)^\top \otimes \frac{\partial \mathcal{L}}{\partial \hat{z}^{(k)}}, \quad \frac{\partial \mathcal{L}}{\partial \hat{z}^{(k)}} = \left[H\left(\hat{z}_L^{(k)}\right), H\left(\hat{z}_U^{(k)}\right)\right] \odot \frac{\partial \mathcal{L}}{\partial z^{(k)}}$$

$$\frac{\partial \mathcal{L}}{\partial W^{(k)}} = \frac{\partial \mathcal{L}}{\partial \hat{z}^{(k)}} \otimes \left[\left(z_L^{(k-1)}\right)^\top, \left(z_U^{(k-1)}\right)^\top\right], \quad \frac{\partial \mathcal{L}}{\partial b^{(k)}} = \frac{\partial \mathcal{L}}{\partial \hat{z}^{(k)}}$$

where $H(\cdot)$ is the Heaviside function (assuming we are using a rectified linear activation function, otherwise see [13] for other choices), and ∘ is the Hadamard product. The resulting intervals are valid bounds for each partial derivative, that is $\partial \mathcal{L}/\partial W^{(k)} \in \partial \mathcal{L}/\partial W^{(k)}$ for all $W^{(k)} \in W^{(k)}$, $\partial l/\partial y^\star \in [\partial l_L, \partial l_U]$ and $\hat{z}^{(k)} \in \left[\hat{z}_L^{(k)}, \hat{z}_U^{(k)}\right]$.

4.7.3 Discussion of Certification

In many ways, certification inverts the strengths and drawbacks of the sampling procedure. In particular, computing valid gradient bounds requires that we have complete knowledge of the models architecture and weights, fails to scale to large models/deep neural networks (beyond 3 layers), and can vastly underestimate the robustness of the models we investigate. The great strength of these methods, in contrast to all other methods we have discussed, is the ability of valid gradient bounds to formally guarantee that the explanation is robust. In this way, the method of certification serves as a strong compliment to the other methods we have discussed. Indeed, no one method on its own is able to deliver a comprehensive analysis of the robustness of a machine learning model.

4.8 Literature Review

Though it is outside of the scope of this work to provide a comprehensive literature review of the works that study robustness of saliency-based explanations, in this section we touch on the seminal works related to the concepts we have covered in this chapter.

Related Evaluations in Explainability In this book we focus on the robustness of explanations; we establish that without this core desideratum explanations cannot be useful. In our discussion we rarely discuss how one might evaluate an explanations utility for its end-goals (e.g., engendering end-user trust, debugging, fairness evaluations). Largely, this is outside the aims of this book, however, here we provide interested readers with some relevant literature that discusses this topic. In [1], user surveys are conducted which establish that saliency-based explanations are primarily used by model developers for debugging purposes and it identifies opportunities and goals for broader adoption. Further in [22], the authors conduct workshops with end-users to clarify how explanations can be improved to meet end-users needs. In [23], the authors evaluate the impact of different kinds of explanation on user trust and understanding of model predictions. Each of these works make take a critical step (user-studies) toward deployment of explainability in real-world settings.

Adversarial Attacks Adversarial examples, inputs that have been imperceptibly modified to induce misclassification (here, not to fool explanations), are a well-known vulnerability of neural networks [14, 24, 25]. A significant amount of research has studied methods for proving that no adversary can change the prediction for a given input and perturbation budget [26–28]. Analogous attacks on gradient-based explanations have been explored [6, 8, 11, 12]. In [11, 12] the authors investigate how to maximally perturb the explanation using first-order methods similar to the projected gradient descent attack on predictions [25]. In [6–8, 10] the authors investigate perturbing model parameters rather than inputs with the goal of finding a model that globally produces corrupted explanations while maintaining model performance. This attack on explanations has been extended to a worrying use-case of disguising model bias [10].

Enhancing Robustness During Training Methods that seek to remedy the lack of robustness work by either modifying the training procedure [29, 30] or by modifying the explanation method [9, 30, 31]. Methods that modify the model include normalisation of the Hessian norm during training in order to give a penalty on principle curvatures [29, 30]. Further work suggests attributional adversarial training, searching for the points that maximise the distance between local gradients [32–35]. Works that seek to improve explanation in a model agnostic way include smoothing the gradient by adding random noise [2, 36] or by using an ensemble of explanation methods [37]. The above defenses cannot rule out potential success of more sophisticated adversaries and it is well-known that the approaches proposed for improving the robustness of explanations do not work against adaptive adversaries in the case of prediction robustness [38, 39]. However, *certification* methods have emerged which provide sound guarantees that even the most powerful adversary cannot fool a models prediction [40–42]. Moreover, these methods have been made differentiable and used for training models with provable prediction robustness and for small networks achieve state-of-the-art robustness results [19, 43, 44]. This work extends these certification methods to the problem of robust explanations. By adapting results from symbolic interval analysis of Bayesian neural networks [45, 46] and applying them to interval analysis of both the forward pass *and* backwards pass we are able to get guarantees on the robustness of gradient-based explanations.

Robust-by-Design Explanations A key avenue of research id designing explanation methods that use robustness as a primary desiderata. In [47] the authors rely on abductive reasoning to get guarantees that are guaranteed to be *minimal* (e.g., in the number of explaining features used) but are not guaranteed to be robust. The authors of [48] build on abductive-based explanations and consider explanations that are both minimal and optimally robust. In [49] the authors also consider provable robustness as a key desideratum and achieve this by sampling a black-box model exponentially many times and thus deriving statistical guarantees. Similarly an iterative, greedy strategy is employed by [50] to get some statistical guarantees on the quality of their explanations.In [51], the authors provide provable guarantees in terms of optimal distance, i.e., nearest explanation, and perfect coverage.

4.9 Discussion and Future Work

In this section, we discuss several important directions of future work. Several questions we raise in this section mirror the questions discussed in our previous chapter on counterfactuals, though here we focus on these questions in the context of saliency-based explanations and their robustness.

Research Question 1: Further Robustness Evaluations In this chapter we have covered the primary goals of robustness evaluation for saliency-based explanation including evaluation with noise, audits with attacks, and certification; however, we have only presented the most straight-forward computational approaches in each category. How to meaningfully extend these approaches to discrete settings such as natural language processing is an open question. Moreover, independent advances in fields such as optimisation may prove impact when applied to robustness evaluation for saliency-based explanations. One such example would be applying the latest advances in bound propagation to improve what we have discussed in Sect. 4.7 of this chapter.

Research Question 2: Incorporating Explanations into Training One key component not touched on in this work, but discussed in some detail e.g., in [52] and [53] is the incorporation of robust explanation constraints during the training of a machine learning model. In particular, for any of the evaluation metrics we have discussed above, one can append this metric onto the loss function used at training as a regularisation term. While this may make finding a suitable classifier more difficult, future works into exactly how and when this regularisation should be applied will be important to ensuring saliency-based explanations meet expected robustness criteria.

Research Question 3: Common Benchmarks for Evaluation Developing unified benchmarks for evaluating the robustness of saliency-based explanations will be crucial to advancing the study of robustness in explanations. In particular, benchmark datasets should include specified perturbation scenarios to ensure consistent and comparable robustness evaluations across models and explanation methods. Moreover, they should encompass common pitfalls such as shortcut learning, bias, and spurious correlations–issues that saliency explanations must robustly detect and account for in order to provide utility to all stakeholders. By integrating these challenges into benchmarks, researchers can more effectively assess how well explanation methods handle real-world complexities and provide reliable insights, ultimately driving progress in developing more trustworthy machine learning models.

Research Question 4: Human Evaluations of Robustness Explanations and the robustness criteria they must satisfy, especially given that each application domain comes with its own stake-holder needs and expectations. While robustness is a prerequisite to meaningful explanations, it alone does not capture how explanations are interpreted by end users. Future user studies will help reveal which types of robustness failures, such as sensitivity to input or model noise most undermine the trust in explanations. Involving diverse users, from experts to decision-makers, ensures that explanations are not only technically robust but also comprehensible and useful in practice, making them more valuable for real-world decision-making.

References

1. Umang Bhatt, Alice Xiang, Shubham Sharma, Adrian Weller, Ankur Taly, Yunhan Jia, Joydeep Ghosh, Ruchir Puri, José MF Moura, and Peter Eckersley. Explainable machine learning in deployment. In *Proceedings of the 2020 conference on fairness, accountability, and transparency*, pages 648–657, 2020.
2. Mukund Sundararajan, Ankur Taly, and Qiqi Yan. Axiomatic attribution for deep networks. In *International conference on machine learning*, pages 3319–3328. PMLR, 2017.
3. Marco Túlio Ribeiro, Sameer Singh, and Carlos Guestrin. "why should I trust you?": Explaining the predictions of any classifier. In Balaji Krishnapuram, Mohak Shah, Alexander J. Smola, Charu C. Aggarwal, Dou Shen, and Rajeev Rastogi, editors, *Proceedings of the 22nd ACM SIGKDD International Conference on Knowledge Discovery and Data Mining, San Francisco, CA, USA, August 13–17, 2016*, pages 1135–1144. ACM, 2016.
4. Steven Bramhall, Hayley Horn, Michael Tieu, and Nibhrat Lohia. Qlime-a quadratic local interpretable model-agnostic explanation approach. *SMU Data Science Review*, 3(1):4, 2020.
5. Saumitra Mishra, Bob L Sturm, and Simon Dixon. Local interpretable model-agnostic explanations for music content analysis. In *ISMIR*, volume 53, pages 537–543, 2017.
6. Juyeon Heo, Sunghwan Joo, and Taesup Moon. Fooling neural network interpretations via adversarial model manipulation. In *NeurIPS*, 2019.
7. Dylan Slack, Sophie Hilgard, Emily Jia, Sameer Singh, and Himabindu Lakkaraju. Fooling lime and shap: Adversarial attacks on post hoc explanation methods. In *AAAI*, 2020.
8. Himabindu Lakkaraju and Osbert Bastani. "How do I fool you?" manipulating user trust via misleading black box explanations. In *Proceedings of the AAAI/ACM Conference on AI, Ethics, and Society*, pages 79–85, 2020.
9. Christopher Anders, Plamen Pasliev, Ann-Kathrin Dombrowski, Klaus-Robert Müller, and Pan Kessel. Fairwashing explanations with off-manifold detergent. In *International Conference on Machine Learning*, pages 314–323. PMLR, 2020.
10. Botty Dimanov, Umang Bhatt, Mateja Jamnik, and Adrian Weller. You shouldn't trust me: Learning models which conceal unfairness from multiple explanation methods. In *ECAI*, 2020.

References

11. Ann-Kathrin Dombrowski, Maximillian Alber, Christopher Anders, Marcel Ackermann, Klaus-Robert Müller, and Pan Kessel. Explanations can be manipulated and geometry is to blame. *Advances in Neural Information Processing Systems*, 32, 2019.
12. Amirata Ghorbani, Abubakar Abid, and James Zou. Interpretation of neural networks is fragile. In *AAAI*, 2019.
13. Matthew Robert Wicker, Juyeon Heo, Luca Costabello, and Adrian Weller. Robust explanation constraints for neural networks. In *The Eleventh International Conference on Learning Representations*.
14. Ian J Goodfellow, Jonathon Shlens, and Christian Szegedy. Explaining and harnessing adversarial examples. In *ICLR*, 2015.
15. Christopher Anders, Plamen Pasliev, Ann-Kathrin Dombrowski, Klaus-Robert Müller, and Pan Kessel. Fairwashing explanations with off-manifold detergent. In *International Conference on Machine Learning*, pages 314–323. PMLR, 2020.
16. Bolei Zhou, Aditya Khosla, Agata Lapedriza, Aude Oliva, and Antonio Torralba. Learning deep features for discriminative localization. In *Proceedings of the IEEE conference on computer vision and pattern recognition*, pages 2921–2929, 2016.
17. Ramprasaath R Selvaraju, Michael Cogswell, Abhishek Das, Ramakrishna Vedantam, Devi Parikh, and Dhruv Batra. Grad-cam: Visual explanations from deep networks via gradient-based localization. In *Proceedings of the IEEE international conference on computer vision*, pages 618–626, 2017.
18. Nicholas Carlini and David Wagner. Towards evaluating the robustness of neural networks. In *2017 IEEE symposium on security and privacy (sp)*, pages 39–57. IEEE, 2017.
19. Matthew Mirman, Timon Gehr, and Martin Vechev. Differentiable abstract interpretation for provably robust neural networks. In *International Conference on Machine Learning*, pages 3578–3586. PMLR, 2018.
20. Siegfried M Rump. Fast and parallel interval arithmetic. *BIT Numerical Mathematics*, 39(3):534–554, 1999.
21. Nguyen Hong Diep. Efficient implementation of interval matrix multiplication. In *Applied Parallel and Scientific Computing: 10th International Conference, PARA 2010, Reykjavík, Iceland, June 6–9, 2010, Revised Selected Papers, Part II 10*, pages 179–188. Springer, 2012.
22. Umang Bhatt, McKane Andrus, Adrian Weller, and Alice Xiang. Machine learning explainability for external stakeholders. *arXiv preprint arXiv:2007.05408*, 2020.
23. Jasper van der Waa, Elisabeth Nieuwburg, Anita Cremers, and Mark Neerincx. Evaluating xai: A comparison of rule-based and example-based explanations. *Artificial intelligence*, 291:103404, 2021.
24. Christian Szegedy, Wojciech Zaremba, Ilya Sutskever, Joan Bruna, Dumitru Erhan, Ian Goodfellow, and Rob Fergus. Intriguing properties of neural networks. *arXiv e-prints*, page arXiv:1312.6199, December 2013.
25. Aleksander Madry, Aleksandar Makelov, Ludwig Schmidt, Dimitris Tsipras, and Adrian Vladu. Towards deep learning models resistant to adversarial attacks. In *ICLR*, 2018.
26. Vincent Tjeng, Kai Xiao, and Russ Tedrake. Evaluating robustness of neural networks with mixed integer programming. *arXiv preprint arXiv:1711.07356*, 2017.
27. Tsui-Wei Weng, Huan Zhang, Hongge Chen, Zhao Song, Cho-Jui Hsieh, Duane Boning, Inderjit S Dhillon, and Luca Daniel. Towards fast computation of certified robustness for relu networks. *arXiv preprint arXiv:1804.09699*, 2018.
28. Mahyar Fazlyab, Manfred Morari, and George J Pappas. Safety verification and robustness analysis of neural networks via quadratic constraints and semidefinite programming. *arXiv preprint arXiv:1903.01287*, 2019.
29. Ann-Kathrin Dombrowski, Christopher J Anders, Klaus-Robert Müller, and Pan Kessel. Towards robust explanations for deep neural networks. *Pattern Recognition*, 121:108194, 2022.
30. Zifan Wang, Haofan Wang, Shakul Ramkumar, Matt Fredrikson, Piotr Mardziel, and Anupam Datta. Smoothed geometry for robust attribution. In *NeurIPS Workshops*, 2020.

31. Nianwen Si, Heyu Chang, and Yichen Li. A simple and effective method to defend against saliency map attack. In *International Conference on Frontiers of Electronics, Information and Computation Technologies*, pages 1–8, 2021.
32. Jiefeng Chen, Xi Wu, Vaibhav Rastogi, Yingyu Liang, and Somesh Jha. Robust attribution regularization. *Advances in Neural Information Processing Systems*, 32, 2019.
33. Adam Ivankay, Ivan Girardi, Chiara Marchiori, and Pascal Frossard. Far: A general framework for attributional robustness. *arXiv preprint arXiv:2010.07393*, 2020.
34. Mayank Singh, Nupur Kumari, Puneet Mangla, Abhishek Sinha, Vineeth N Balasubramanian, and Balaji Krishnamurthy. Attributional robustness training using input-gradient spatial alignment. In *European Conference on Computer Vision*, pages 515–533. Springer, 2020.
35. Himabindu Lakkaraju, Nino Arsov, and Osbert Bastani. Robust and stable black box explanations. In *International Conference on Machine Learning*, pages 5628–5638. PMLR, 2020.
36. Daniel Smilkov, Nikhil Thorat, Been Kim, Fernanda Viégas, and Martin Wattenberg. Smoothgrad: removing noise by adding noise. *arXiv preprint arXiv:1706.03825*, 2017.
37. Laura Rieger and Lars Kai Hansen. A simple defense against adversarial attacks on heatmap explanations. *arXiv preprint arXiv:2007.06381*, 2020.
38. Anish Athalye, Nicholas Carlini, and David Wagner. Obfuscated gradients give a false sense of security: Circumventing defenses to adversarial examples. In *International conference on machine learning*, pages 274–283. PMLR, 2018.
39. Warren He, James Wei, Xinyun Chen, Nicholas Carlini, and Dawn Song. Adversarial example defense: Ensembles of weak defenses are not strong. In *11th USENIX workshop on offensive technologies (WOOT 17)*, 2017.
40. Matthew Wicker, Xiaowei Huang, and Marta Kwiatkowska. Feature-guided black-box safety testing of deep neural networks. In *International Conference on Tools and Algorithms for the Construction and Analysis of Systems*, pages 408–426. Springer, 2018.
41. Timon Gehr, Matthew Mirman, Dana Drachsler-Cohen, Petar Tsankov, Swarat Chaudhuri, and Martin Vechev. Ai2: Safety and robustness certification of neural networks with abstract interpretation. In *2018 IEEE S&P*, pages 3–18. IEEE, 2018.
42. Min Wu, Matthew Wicker, Wenjie Ruan, Xiaowei Huang, and Marta Kwiatkowska. A game-based approximate verification of deep neural networks with provable guarantees. *Theoretical Computer Science*, 807:298–329, 2020.
43. Sven Gowal, Krishnamurthy Dvijotham, Robert Stanforth, Rudy Bunel, Chongli Qin, Jonathan Uesato, Relja Arandjelovic, Timothy Mann, and Pushmeet Kohli. On the effectiveness of interval bound propagation for training verifiably robust models. *arXiv preprint arXiv:1810.12715*, 2018.
44. Matthew Wicker, Luca Laurenti, Andrea Patane, Zhuotong Chen, Zheng Zhang, and Marta Kwiatkowska. Bayesian inference with certifiable adversarial robustness. In *International Conference on Artificial Intelligence and Statistics*, pages 2431–2439. PMLR, 2021.
45. Matthew Wicker, Luca Laurenti, Andrea Patane, and Marta Kwiatkowska. Probabilistic safety for bayesian neural networks. In *Conference on Uncertainty in Artificial Intelligence*, pages 1198–1207. PMLR, 2020.
46. Leonard Berrada, Sumanth Dathathri, Krishnamurthy Dvijotham, Robert Stanforth, Rudy R Bunel, Jonathan Uesato, Sven Gowal, and M Pawan Kumar. Make sure you're unsure: A framework for verifying probabilistic specifications. *Advances in Neural Information Processing Systems*, 34:11136–11147, 2021.
47. Alexey Ignatiev, Nina Narodytska, and Joao Marques-Silva. Abduction-based explanations for machine learning models. In *Proceedings of the AAAI Conference on Artificial Intelligence*, volume 33, pages 1511–1519, 2019.
48. Emanuele La Malfa, Rhiannon Michelmore, Agnieszka M. Zbrzezny, Nicola Paoletti, and Marta Kwiatkowska. On guaranteed optimal robust explanations for nlp models. In *Proceedings of the Thirtieth International Joint Conference on Artificial Intelligence, IJCAI-21*, 2021.

References

49. Guy Blanc, Jane Lange, and Li-Yang Tan. Provably efficient, succinct, and precise explanations. *Advances in Neural Information Processing Systems*, 34:6129–6141, 2021.
50. Marco Tulio Ribeiro, Sameer Singh, and Carlos Guestrin. Anchors: High-precision model-agnostic explanations. In *Proceedings of the AAAI conference on artificial intelligence*, volume 32, 2018.
51. Kiarash Mohammadi, Amir-Hossein Karimi, Gilles Barthe, and Isabel Valera. Scaling guarantees for nearest counterfactual explanations. In *Proceedings of the 2021 AAAI/ACM Conference on AI, Ethics, and Society*, pages 177–187, 2021.
52. Andrew Slavin Ross, Michael C Hughes, and Finale Doshi-Velez. Right for the right reasons: Training differentiable models by constraining their explanations. *arXiv preprint arXiv:1703.03717*, 2017.
53. Juyeon Heo, Vihari Piratla, Matthew Wicker, and Adrian Weller. Use perturbations when learning from explanations. *Advances in Neural Information Processing Systems*, 36:26872–26897, 2023.

The manufacturer's authorised representative in the EU is Springer Nature Customer Service Centre GmbH, Europaplatz 3, 69115 Heidelberg, Germany. If you have any concerns regarding our products, please contact ProductSafety@springernature.com

Printed and bound by CPI Group (UK) Ltd, Croydon, CR0 4YY

26/03/2026

02078953-0012